SpringerBriefs in Statistics

For further volumes:
http://www.springer.com/series/8921

Shigeru Yamada

Software Reliability Modeling

Fundamentals and Applications

 Springer

Shigeru Yamada
Graduate School of Engineering
Tottori University
Tottori
Japan

ISSN 2191-544X   ISSN 2191-5458 (electronic)
ISBN 978-4-431-54564-4   ISBN 978-4-431-54565-1 (eBook)
DOI 10.1007/978-4-431-54565-1
Springer Tokyo Heidelberg New York Dordrecht London

Library of Congress Control Number: 2013950356

Printed on acid-free paper

Springer is part of Springer Science+Business Media (www.springer.com)

Preface

Software reliability is one of the most important characteristics of software product quality. Its measurement and management technologies during the software product life-cycle are essential to produce and maintain quality/reliable software systems.

Chapter 1 of this book introduces several aspects of software reliability modeling and its applications. Hazard rate and nonhomogeneous Poisson process (NHPP) models are investigated particularly for quantitative software reliability assessment. Further, imperfect debugging and software availability models are discussed with reference to incorporating practical factors of dynamic software behavior. Three software management problems are presented as application technologies of software reliability models: the optimal software release problem, the statistical testing-process control, and the optimal testing-effort allocation problem.

Chapter 2 of this book describes several recent developments in software reliability modeling and their applications as quantitative techniques for software quality/reliability measurement and assessment. The discussion includes a quality engineering analysis of human factors affecting software reliability during the design review phase, which is the upper stream of software development, as well as software reliability growth models based on stochastic differential equations and discrete calculus during the testing phase, which is the lower stream. From the point of view of quantitative project management, quality-oriented software management analysis by applying the multivariate analysis method and the existing software reliability growth models to actual process monitoring data are also discussed. The final part of the Chap. 2 provides an illustration of operational performability evaluation for the software-based system, by introducing the concept of systemability defined as the reliability characteristic subject to the uncertainty of the field environment.

Tottori, Japan, May 2013   Shigeru Yamada

v

Acknowledgments

I would like to express my sincere appreciation to Drs. Shinji Inoue, Yoshinobu Tamura, and Koichi Tokuno for their helpful suggestions in completion of this monograph on software reliability modeling. Thanks also go to my research colleagues from universities and industry for their warm advice. I am also indebted to Mr. Takahiro Nishikawa, Department of Social Management Engineering, Graduate School of Engineering, Tottori University, Japan, for his support in the editing.

Contents

Chapter 1
Introduction to Software Reliability Modeling and Its Applications

Abstract Software reliability is one of the most important characteristics of software quality. Its measurement and management technologies during the software life-cycle are essential to produce and maintain quality/reliable software systems. In this chapter, we discuss software reliability modeling and its applications. As to software reliability modeling, hazard rate and NHPP models are investigated particularly for quantitative software reliability assessment. Further, imperfect debugging and software availability models are also discussed with reference to incorporating practical factors of dynamic software behavior. And three software management problems are discussed as an application technology of software reliability models: the optimal software release problem, statistical testing-progress control, and the optimal testing-effort allocation problem.

Keywords Software product quality/reliability assessment · Software reliability growth modeling · Nonhomogeneous Poisson process · Imperfect debugging · Software availability · Markov process · Optimal release problem · Testing-progress control · Testing-effort allocation

1.1 Introduction

In recent years, many computer system failures have been caused by software faults introduced during the software development process. This is an inevitable problem, since an software system installed in the computer system is an intellectual product consisting of documents and source programs developed by human activities. Then, total quality management (TQM) is considered to be one of the key technologies needed to produce more highly quality software products [1, 2]. In the case of TQM used for software development, all phases of the development process, i.e. requirement specification, design, coding, and testing, have to be controlled systematically to prevent the introduction of software bugs or faults as far as possible and to

S. Yamada, *Software Reliability Modeling*, SpringerBriefs in Statistics, DOI: 10.1007/978-4-431-54565-1_1, © The Author(s) 2014

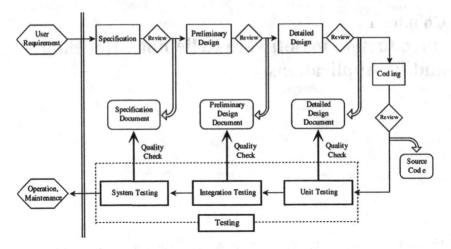

Fig. 1.1 A general software development process (water-fall paradigm)

detect any introduced faults in the software system as early as possible. Basically, the concept of TQM means assuring the quality of the products in each phase to the next phase. Particularly, quality control carried out at the testing phase, which is the last stage of the software development process, is very important. During the testing phase, the product quality and the software performance during the operation phase are evaluated and assured. In concrete terms, a lot of software faults introduced in the software system through the first three phases of the development process by human activities are detected, corrected, and removed. Figure 1.1 shows a general software development process called a waterfall paradigm.

Therefore, TQM for software development, i.e. *software TQM*, has been emphasized. Software TQM aims to manage the software life-cycle comprehensively, considering productivity, quality, cost and delivery simultaneously, and assure software quality elements in Fig. 1.2. In particular, the management technologies for improving software reliability are very important. The quality characteristic of software reliability is that computer systems can continue to operate regularly without the occurrence of failures in software systems.

In this chapter, we discuss a quantitative technique for software quality/reliability measurement and assessment (see Fig. 1.3) as one of the key software reliability technologies, which is a so-called *software reliability model* (abbreviated as *SRM*), and its applications.

1.2 Definitions and Software Reliability Model

Generally, a software failure caused by software faults latent in the system cannot occur except on a specific occasion when a set of specific data is put into the system under a specific condition, i.e. the program path including software faults is executed.

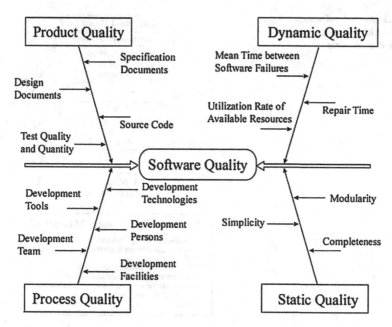

Fig. 1.2 Elements of software quality based on a cause-and-effect diagram

Therefore, the software reliability is dependent on the input data and the internal condition of the program. We summarize the definitions of the technical terms related to the software reliability below.

A *software system* is a product which consists of the programs and documents produced through the software development process discussed in the previous section (see Fig. 1.1). The specification derived by analyzing user requirements for the software system is a document which describes the expected performance and function of the system. When the software performance deviates from the specification and an output variable has an improper value or the normal processing is interrupted, it is said that a software failure occurs. That is, *software failure* is defined as an unacceptable departure of program operation from the program requirements. The cause of software failure is called a software fault. Then, *software fault* is defined as a defect in the program which causes a software failure. The software fault is usually called a *software bug*. *Software error* is defined as human action that results in the software system containing a software fault [3, 4]. Thus, the software fault is considered to be a manifestation of software errors.

Based on the basic definitions above, we can describe a software behavior as Input(I)-Program(P)-Output(O) model [5, 6], as shown in Fig. 1.4.

In this model a program is considered as a mapping from the input space constituting input data available on use to the output space constituting output data or interruptions of normal processing. Testing space T is an input subspace of I, the performance of which can be verified and validated by software testing. Software

Fig. 1.3 Aim of software quality/reliability measurement and assessment

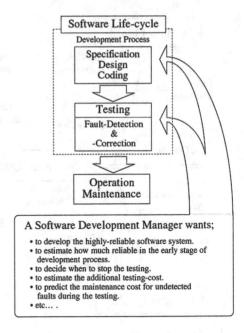

faults detected and removed during the testing phase map the elements of input subspace E into an output subspace O' constituting the events of a software failure. That is, the faults detected during the testing phase belong to the intersection of subspace E and T. Software faults remaining in the operation phase belong to the subspace E but not to the testing space T.

Figure 1.5 compares the characteristics between software reliability and hardware reliability. Under the definitions for technical terms above, *software reliability* is defined as the attribute that a software system will perform without causing software failures over a given time period under specified conditions, and is measured by its probability [3, 4]. A *software reliability model* (*SRM*) is a mathematical analysis model for the purpose of measuring and assessing software quality/reliability quantitatively. Many software reliability models have been proposed and applied to practical use because software reliability is considered to be a "*must-be quality* "characteristic of a software product. The software reliability models can be divided into two classes [6, 7] as shown in Fig. 1.6. One treats the upper software development process, i.e. design and coding phases, and analyzes the reliability factors of the software products and processes, which is categorized in the class of static model. The other deals with testing and operation phases by describing a software failure-occurrence phenomenon or software fault-detection phenomenon, by applying the stochastic/statistics theories and can estimate and predict the software reliability, which is categorized in dynamic model.

In the former class, a *software complexity model* is well known and can measure the reliability by assessing the complexity based the structural characteristics

Terminologies

- *Software Failure*
  — an unacceptable departure of program operation from the program requirements
- *Software Fault*
  — a defect in the program which causes a software failure (*software bug*)
- *Software Reliability*
  — the attribute that a software system will perform without causing software failures over a given time period under specified conditions
  — measured by its probability

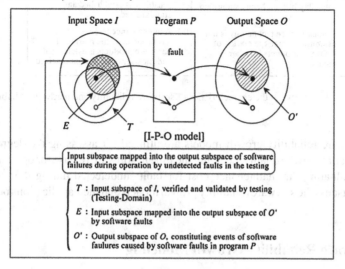

Fig. 1.4 An input-program-output model for software behavior

of products and the process features to produce the products. In the latter class, a *software reliability growth model* is especially well known. Further, this model is divided into three categories [6, 7]:

(1) *Software failure-occurrence time model*
The model which is based on the software failure-occurrence time or the software fault-detection time.
(2) *Software fault-detection count model*
The model which is based on the number of software failure-occurrences or the number of detected faults.
(3) *Software availability model*
The model which describes the time-dependent behavior of software system alternating up (operation) and down (restoration or fault correction) states.

Software Reliability	Hardware Reliability
(1) Software failures can be due to no wearout phenomenon.	(1) Hardware failures can be due to wear.
(2) Software reliability is, inherently, determined during the earlier phase of the development process, i.e. specification analysis and design phases.	(2) Hardware reliability is affected by deficiencies injected during all phases of the development, operation, and maintenance.
(3) Software reliability can not be improved by redundancy with identical versions.	(3) Hardware reliability can be improved by redundancy with identical units.
(4) A verification method of software reliability has not been established.	(4) A testing method of hardware reliability is established and standardized.
(5) A maintenance technology is not established since the market of software products is rather recent.	(5) A maintenance technology is advanced since the market of hardware products is established and the user environment is seized.

Fig. 1.5 Comparison between the characteristics of software reliability and hardware reliability

The software reliability growth models are utilized for assessing the degree of achievement of software quality, deciding the time to software release for operational use, and evaluating the maintenance cost for faults undetected during the testing phase. We discuss the software reliability growth models and their applications below.

1.3 Software Reliability Growth Modeling

Generally, a mathematical model based on stochastic and statistical theories is useful to describe the software fault-detection phenomena or the software failure-occurrence phenomena and estimate the software reliability quantitatively. During the testing phase in the software development process, software faults are detected and removed with a lot of testing-effort expenditures. Then, the number of faults remaining in the software system decreases as the testing goes on. This means that the probability of software failure-occurrence is decreasing, so that the software reliability is increasing and the time-interval between software failures becoming longer with the testing time (see Fig. 1.7).

A mathematical tool which describes software reliability aspect is a *software reliability growth model* [6–9].

Based on the definitions discussed in the previous section, we can develop a software reliability growth model based on the assumptions used for the actual environment during the testing phase or the operation phase. Then, we can define the following random variables on the number of detected faults and the software failure-occurrence time (see Fig. 1.8):

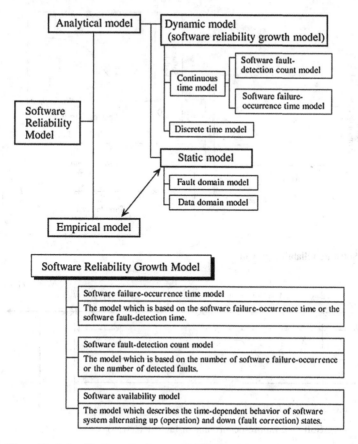

Fig. 1.6 Hierarchical classification of software reliability models

$N(t)$ the cumulative number of software faults (or the cumulative number of observed software failures) detected up to time t,

S_i the ith software-failure occurrence time ($i = 1, 2, \ldots$; $S_0 = 0$),

X_i the time-interval between $(i-1)$-st and ith software failures ($i = 1, 2, \ldots$; $X_0 = 0$).

Figure 1.8 shows the occurrence of event $\{N(t) = i\}$ since i faults have been detected up to time t. From these definitions, we have

$$S_i = \sum_{k=1}^{i} X_k, \qquad X_i = S_i - S_{i-1}. \tag{1.1}$$

Assuming that the *hazard rate*, i.e. the *software failure rate*, for $X_i (i = 1, 2, \ldots)$, $z_i(x)$, is proportional to the current number of residual faults remaining in the system, we have

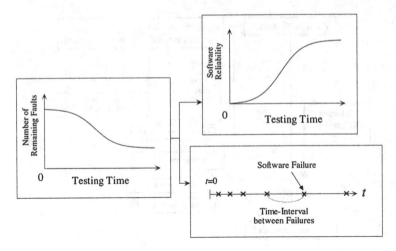

Fig. 1.7 Software reliability growth

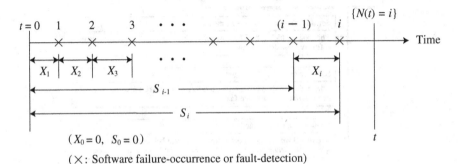

(×: Software failure-occurrence or fault-detection)

Fig. 1.8 The stochastic quantities related to a software fault-detection phenomenon or a software failure-occurrence phenomenon

$$z_i(x) = (N - i + 1)\lambda(x), \quad i = 1, 2, \ldots, N; x \geq 0, \lambda(x) > 0, \quad (1.2)$$

where N is the initial fault content and $\lambda(x)$ the software failure rate per fault remaining in the system at time x. If we consider two special cases in (1.2) as

$$\lambda(x) = \phi, \quad \phi > 0, \quad (1.3)$$

$$\lambda(x) = \phi x^{m-1}, \quad \phi > 0, m > 0, \quad (1.4)$$

then two typical software hazard rate models, respectively called the Jelinski-Moranda model [10] and the Wagoner model [11] can be derived, where ϕ and m are constant parameters. Usually, it is difficult to assume that a software system is completely fault free or failure free. Then, we have a software hazard rate model called the Moranda model [12] for the case of the infinite number of software failure occurrences as

$$z_i(x) = Dk^{i-1}, \quad i = 1, 2, \ldots; D > 0, 0 < k < 1, \tag{1.5}$$

where D is the initial software hazard rate and k the decreasing ratio. Equation (1.5) describes a software failure-occurrence phenomenon where a software system has high frequency of software failure occurrence during the early stage of the testing or the operation phase and it gradually decreases thereafter (see Fig. 1.9). Based on the software hazard rate models above, we can derive several software reliability assessment measures (see Fig. 1.10). For example, the *software reliability* function for $X_i (i = 1, 2, \ldots)$ is given as

$$R_i(x) = \exp\left[-\int_0^x z_i(x)dx\right], \quad i = 1, 2, \ldots. \tag{1.6}$$

Further, we also discuss NHPP models [8, 13–15], which are modeled for random variable $N(t)$ as typical software reliability growth models (see Fig. 1.11). In

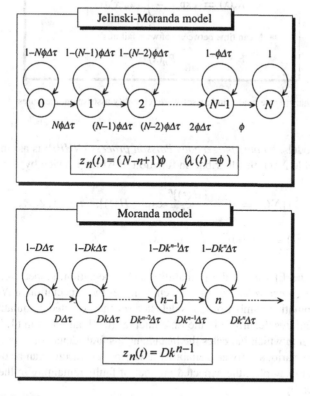

Fig. 1.9 The Jelinski-Moranda model and Moranda model

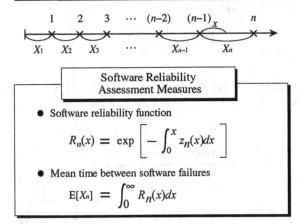

Fig. 1.10 The hazard rate model and its software reliability assessment measures

the NHPP models, a *nonhomogeneous Poisson process (NHPP)* is assumed for the random variable $N(t)$, the distribution function of which is given by

$$\Pr\{N(t) = n\} = \frac{\{H(t)\}^n}{n!} \exp[-H(t)], \quad n = 1, 2, \ldots,$$

$$H(t) \equiv E[N(t)] = \int_0^t h(x)dx, \tag{1.7}$$

where $\Pr\{\cdot\}$ and $E[\cdot]$ mean the probability and expectation, respectively. $H(t)$ in (1.7) is called a *mean value function* which indicates the expectation of $N(t)$, i.e. the expected cumulative number of faults detected (or the expected cumulative number of software failures occurred) in the time interval $(0, t]$, and $h(t)$ in (1.7) called an *intensity function* which indicates the instantaneous fault-detection rate at time t.

From (1.7), various software reliability assessment measures can be derived (see Fig. 1.12). For examples, the expected number of faults remaining in the system at time t is given by

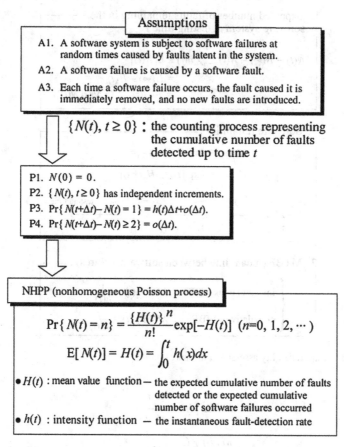

Fig. 1.11 The NHPP model

$$n(t) = a - H(t), \tag{1.8}$$

where $a \equiv H(\infty)$, i.e. parameter a denotes the expected initial fault content in the software system. Given that the testing or the operation has been going on up to time t, the probability that a software failure does not occur in the time-interval $(t, t + x](x \geq 0)$ is given by conditional probability $\Pr\{X_i > x | S_{i-1} = t\}$ as

$$R(x|t) = \exp[H(t) - H(x + t)], \quad t \geq 0, \ x \geq 0. \tag{1.9}$$

$R(x|t)$ in (1.9) is a so-called *software reliability*. Measures of *MTBF* (mean time between software failures or fault detections) can be obtained follows:

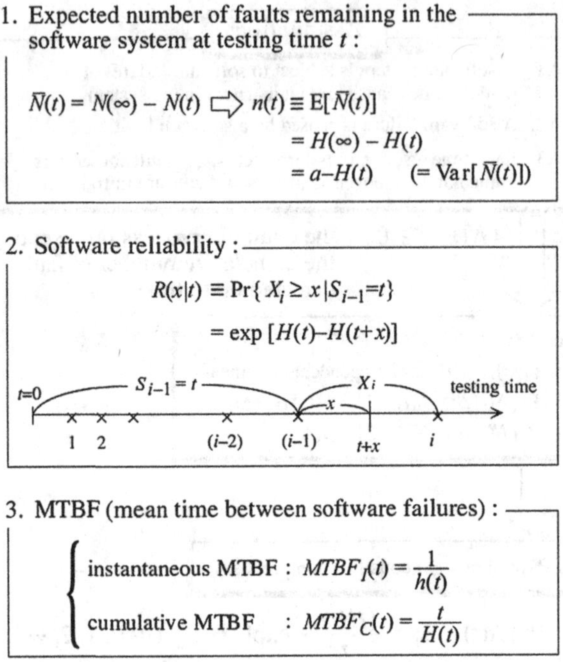

1. **Expected number of faults remaining in the software system at testing time t :**

$$\bar{N}(t) = N(\infty) - N(t) \Rightarrow n(t) \equiv \mathrm{E}[\bar{N}(t)]$$
$$= H(\infty) - H(t)$$
$$= a - H(t) \quad (= \mathrm{Var}[\bar{N}(t)])$$

2. **Software reliability :**

$$R(x|t) \equiv \Pr\{ X_i \ge x | S_{i-1} = t \}$$
$$= \exp [H(t) - H(t+x)]$$

3. **MTBF (mean time between software failures) :**

$$\begin{cases} \text{instantaneous MTBF} : MTBF_I(t) = \dfrac{1}{h(t)} \\[2ex] \text{cumulative MTBF} \quad : MTBF_C(t) = \dfrac{t}{H(t)} \end{cases}$$

Fig. 1.12 Software reliability assessment measures

$$MTBF_I(t) = \frac{1}{h(t)}, \tag{1.10}$$

$$MTBF_C(t) = \frac{t}{h(t)}. \tag{1.11}$$

*MTBF*s in (1.10) and (1.11) are called *instantaneous MTBF* and *cumulative MTBF*, respectively.

It is obvious that the lower the value of $n(t)$ in (1.8), the higher the value $R(x|t)$ for specified x in (1.9), or the longer the value of *MTBF*s in (1.10) and (1.11), the higher the achieved software reliability is. Then, analyzing actual test data with accepted NHPP models, these measures can be utilized to assess software reliability during the testing or operation phase, where statistical inferences, i.e. parameter estimation and goodness-of-fit test, are usually performed by a method of maximum-likelihood.

To assess the software reliability actually, it is necessary to specify the mean value function $H(t)$ in (1.7). Many NHPP models considering the various testing or operation environments for software reliability assessment have been proposed in the last decade [4, 6, 7]. Typical NHPP models are summarized in Table 1.1. As discussed above, a software reliability growth is described as the relationship between the elapsed testing or operation time and the cumulative number of detected faults and can be shown as the reliability growth curve mathematically (see Fig. 1.13).

Table 1.1 A summary of NHPP models

NHPP model	Mean value function $H(t)$	Intensity function $h(t)$	Environment
Exponential software reliability growth model [16, 17]	$m(t) = a(1 - e^{-bt})$ $(a > 0, b > 0)$	$h_m(t) = abe^{-bt}$	A software failure-occurrence phenomenon with a constant fault-detection rate at an arbitrary time is described
Modified exponential software reliability growth model [18, 19]	$m_p(t) = a \sum_{i=1}^{2} p_i(1 - e^{-b_i t})$ $(a > 0, 0 < b_2 < b_1 < 1,$ $\sum_{i=1}^{2} p_i = 1, 0 < p_i < 1)$	$h_p(t) = a \sum_{i=1}^{2} p_i b_i e^{-b_i t}$	A difficulty of software fault-detection during the testing is considered. (b_1 is the fault-detection rate for easily detectable faults; b_2 is the fault-detection rate for hardly detectable faults)
Delayed S-shaped software reliability growth model [20, 21]	$M(t) = a[1 - (1 + bt)e^{-bt}]$ $(a > 0, b > 0)$	$h_M(t) = ab^2 t e^{-bt}$	A software fault-detection process is described by two successive phenomena, i.e. failure-detection process and fault-isolation process
Inflection S-shaped software reliability growth model [22, 23]	$I(t) = \dfrac{a(1 - e^{-bt})}{(1 + ce^{-bt})}$ $(a > 0, b > 0, c > 0)$	$h_I(t) = \dfrac{ab(1 + c)e^{-bt}}{(1 + ce^{-bt})^2}$	A software failure-occurrence phenomenon with mutual dependency of detected faults is described
Testing-effort-dependent software reliability growth model [24, 25]	$T(t) = a[1 - e^{-rW(t)}]$ $W(t) = \alpha(1 - e^{-\beta t^m})$ $(a > 0, 0 < r < 1,$ $\alpha > 0, \beta > 0, m > 0)$	$h_T(t) = ar\alpha\beta$ $\cdot mt^{m-1}e^{-rW(t)}$	The time-dependent behavior of the amount of testing effort and the cumulative number of detected faults is considered

(continued)

Table 1.1 (continued)

Testing-domain-dependent software reliability growth model [26, 27]	$D(t) = a[1 - \dfrac{1}{v-b}(ve^{-bt} - be^{-vt})](v \neq b)$ $h_D(t) = \dfrac{avb}{v-b}(e^{-bt} - e^{-vt})$	The testing domain, which is the set of software functions influenced by executed test cases, is considered
Logarithmic Poisson execution time model [28, 29]	$\mu(t) = \dfrac{1}{\theta}\ln(\lambda_0\theta t + 1)$ $(\lambda_0 > 0, \theta > 0)$ $\lambda(t) = \dfrac{\lambda_0}{(\lambda_0\theta t + 1)}$	When the testing or operation time is measured on the basis of the number of CPU hours, an exponentially decreasing software failure rate is considered with respect to the cumulative number of software failures

a the expected number of initial fault content in the software system
b, b_i, r the parameters representing the fault-detection rate
c the parameter representing the inflection factor of test personnel
p_i the fault content ratio of Type i fault ($i = 1, 2$)
α, β, m the parameters which determine the testing-effort function $W(t)$
v the testing-domain growth rate
λ_0 the initial software failure rate
θ the reduction rate of software failure rate

Among the NHPP models in Table 1.1, exponential and modified exponential soft-ware reliability growth models are appropriate when the observed reliability growth curve shows an exponential curve ((A) in Fig. 1.13). Similarly, delayed S-shaped and inflection S-shaped software reliability growth models are appropriate when the reliability growth curve is S-shaped ((B) in Fig. 1.13).

In addition, as for computer makers or software houses in Japan, *logistic curve* and *Gompertz curve models* have often been used as software quality assessment models, on the assumption that software fault-detection phenomena can be shown by S-shaped reliability growth curves [30, 31]. In these deterministic models, the cumulative number of faults detected up to testing t is formulated by the following growth equations:

$$L(t) = \frac{k}{1 + me^{-\alpha t}}, \quad m > 0, \ \alpha > 0, \ k > 0, \tag{1.12}$$

$$G(t) = ka^{(b^t)}, \quad 0 < a < 1, \ 0 < b < 1, \ k > 0. \tag{1.13}$$

In (1.12) and (1.13), assuming that a convergence value of each curve ($L(\infty)$ or $G(\infty)$), i.e. parameter k, represents the initial fault content in the software system, it can be estimated by a regression analysis.

1.4 Imperfect Debugging Modeling

Most software reliability growth models proposed so far are based on the assump-tion of perfect debugging, i.e. that all faults detected during the testing and operation phases are corrected and removed perfectly. However, debugging actions in real test-ing and operation environment are not always performed perfectly. For example, typing errors invalidate the fault-correction activity or fault-removal is not carried out precisely due to incorrect analysis of test results [32]. We therefore have an inter-est in developing a software reliability growth model which assumes an *imperfect*

Fig. 1.13 Typical software reliability growth curves

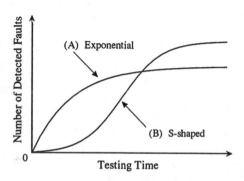

debugging environment (cf. [33, 34]). Such an imperfect debugging model is expected to estimate reliability assessment measures more accurately.

1.4.1 Imperfect Debugging Model with Perfect Correction Rate

To model an imperfect debugging environment, the following assumptions are made:

1. Each fault which causes a software failure is corrected perfectly with probability $p(0 \le p \le 1)$. It is not corrected with probability $q(= 1 - p)$. We call p the perfect debugging rate or the perfect correction rate.
2. The hazard rate is given by (1.5) and decreases geometrically each time a detected fault is corrected (see Fig. 1.14).
3. The probability that two or more software failures occur simultaneously is negligible.
4. No new faults are introduced during the debugging. At most one fault is removed when it is corrected, and the correction time is not considered.

Let $X(t)$ be a random variable representing the cumulative number of faults corrected up to the testing time t. Then, $X(t)$ forms a *Markov process* [35]. That is, from assumption 1, when i faults have been corrected by arbitrary testing time t,

$$X(t) = \begin{cases} i, & \text{with probability } q, \\ i + 1, & \text{with probability } p, \end{cases} \tag{1.14}$$

(see Fig. 1.15). Then, the one-step transition probability for the Markov process that after making a transition into state i, the process $\{X(t), t \ge 0\}$ makes a transition into state j by time t is given by

Fig. 1.14 Behavior of hazard rate

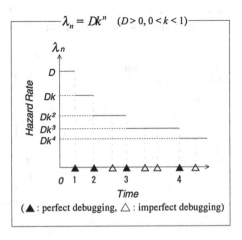

($\blacktriangle$: perfect debugging, $\triangle$: imperfect debugging)

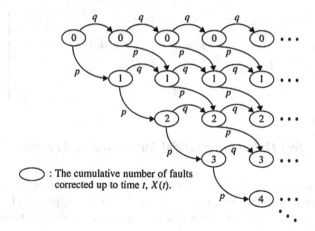

Fig. 1.15 A diagrammatic representation of transitions between states of $X(t)$

$$Q_{ij}(t) = p_{ij}(1 - \exp[-Dk^i t]),\qquad\qquad (1.15)$$

where p_{ij} are the transition probabilities from state i to state j and are given by

$$p_{ij} = \begin{cases} q, & (i = j), \\ p, & (j = i + 1), \quad i, j = 0, 1, 2, \dots. \\ 0, & (\text{elsewhere}), \end{cases}\qquad (1.16)$$

Equation (1.15) represents the probability that if i faults have been corrected at time zero, j faults are corrected by time t after the next software failure occurs. Therefore, based on Markov analysis by using the assumptions and stochastic quantities above, we have the software reliability function and the mean time between software failures for X_i $(i = 1, 2, \dots)$ as

$$R_i(x) = \sum_{s=0}^{i-1} \binom{i-1}{s} p^s q^{i-1-s} \exp[-Dk^s x],\qquad (1.17)$$

$$E[X_i] = \int_0^\infty R_i(x)dx = \frac{(\frac{p}{k} + q)^{i-1}}{D}.\qquad (1.18)$$

And if the initial fault content in the system, N, is specified, the expected cumulative number of faults debugged imperfectly up to time t is given by

$$M(t) = \frac{q}{p} \sum_{n=1}^{N} \sum_{i=0}^{n-1} A_{i,n}(1 - \exp[-pDk^i t]),\qquad (1.19)$$

where $A_{i,n}$ is

$$\left. \begin{aligned} A_{0,1} &\equiv 1 \\ A_{i,n} &= \frac{k^{(1/2)n(n-1)-i}}{\displaystyle\prod_{\substack{j=0 \\ j \neq i}}^{n-1} (k^j - k^i)}, \quad n = 2, 3, \ldots; \ i = 0, 1, 2, \ldots, n-1 \end{aligned} \right\} . \tag{1.20}$$

1.4.2 Imperfect Debugging Model for Introduced Faults

Besides the imperfect debugging factor above in fault-correction activities, we consider the possibility of introducing new faults in the debugging process. It is assumed that the following two kinds of software failures exist in the dynamic environment [36, 37], i.e. the testing or user operation phase:

(F1) software failures caused by faults originally latent in the software system prior to the testing (which are called inherent faults),
(F2) software failures caused by faults introduced during the software operation owing to imperfect debugging.

In addition, it is assumed that one software failure is caused by one fault and that it is impossible to discriminate whether the fault that caused the software failure that has occurred is F1 or F2. As to the software failure-occurrence rate due to F1, the inherent faults are detected with the progress of the operation time. In order to consider two kinds of time dependencies on the decreases of F1, let $a_i(t)(i = 1, 2)$ denote the software failure-occurrence rate for F1. On the other hand, the software failure-occurrence rate due to F2 is denoted as constant $\lambda(\lambda > 0)$, since we assume that F2 occurs randomly throughout the operation. When we consider the software failure-occurrence rate at operation time t is given by

$$h_i(t) = \lambda + a_i(t), \quad i = 1, 2. \tag{1.21}$$

From (1.21), the expected cumulative number of software failures in the time-interval $(0, t]$ (or the expected cumulative number of detected faults) is given by

$$\left. \begin{aligned} H_i(t) &= \lambda t + A_i(t), \\ A_i(t) &= \int_0^t a_i(x)dx, \quad i = 1, 2 \end{aligned} \right\} . \tag{1.22}$$

Then, we have two imperfect debugging models based on an NHPP discussed in Sect. 1.3, where $h_i(t)$ in (1.21) and $H_i(t)$ in (1.22) are used as the intensity functions and the mean value functions ($i = 1, 2$) for an NHPP, respectively. Especially, exponential and delayed S-shaped software reliability growth models are assumed for describing software failure-occurrence phenomena attributable to the inherent faults as (see Table 1.1)

$$a_1(t) = abe^{-bt}, \quad a > 0, b > 0, \tag{1.23}$$
$$a_2(t) = ab^2te^{-bt}, \quad a > 0, b > 0, \tag{1.24}$$

where a is the expected number of initially latent inherent faults and b the software failure-occurrence rate per inherent fault. Therefore, the mean value functions of NHPP models for the imperfect debugging factor are given by

$$H_1(t) = \lambda t + a(1 - e^{-bt}), \tag{1.25}$$
$$H_2(t) = \lambda t + a[1 - (1 + bt)e^{-bt}]. \tag{1.26}$$

From these imperfect debugging models we can derive several software reliability measures for the next software failure-occurrence time interval X since current time t, such as the software reliability function $R_i(x|t)$, the software hazard rate $z_i(x|t)$, and the mean time between software failures $E_i[X|t](i = 1, 2)$:

$$R_i(x|t) = \exp[H_i(t) - H_i(t + x)], \quad t \geq 0, x \geq 0, \tag{1.27}$$

$$z_i(x|t) = -\frac{d}{dx}R_i(x|t)/R_i(x|t) = h_i(t + x), \tag{1.28}$$

$$E_i[X|t] = \int_0^\infty R_i(x|t)dx. \tag{1.29}$$

1.5 Software Availability Modeling

Recently, software performance measures such as the possible utilization factors have begun to be interesting for metrics as well as the hardware products. That is, it is very important to measure and assess *software availability*, which is defined as the probability that the software system is performing successfully, according to the specification, at a specified time point [38–40] (see Fig. 1.16). Several stochastic models have been proposed so far for software availability measurement and assessment. One group [41] has proposed a software availability model considering a reliability growth process, taking account of the cumulative number of corrected faults. Others [42–44] have constructed software availability models describing the uncertainty of fault removal. Still others [45] and [46] have incorporated the increasing difficulty of fault removal.

The actual operational environment needs to be more clearly reflected in software availability modeling, since software availability is a customer-oriented metrics. In [46] and [47] the development of a plausible model is described, which assumes that there exist two types of software failure occurring during the operation phase. Furthermore, in [48] an operational software availability model is built up from the viewpoint of restoration scenarios.

- Software Reliability : ─────────────────

the attribute that software systems can continue to perform and do not cause any software failures *for a given time period*, under a specified environment.

■ *software failure*

[an unacceptable departure from program operation caused by a software fault remaining in the software system.]

- Software Availability : ─────────────────

the attribute that software systems are available and performing *at a given time point*, under a specified environment.

Fig. 1.16 Comparison between software reliability and availability

The above models have employed Markov processes for describing the stochastic time-dependent behaviors of the systems which alternate between the up state, operating regularly, and the restoration state (down state) when a system is inoperable [49]. Several stochastic metrics for software availability measurement in dynamic environment are derived from the respective models.

We discuss a fundamental software availability model [44] below.

1.5.1 Model Description

The following assumptions are made for software availability modeling:

1. The software system is unavailable and starts to be restored as soon as a software failure occurs, and the system cannot operate until the restoration action is complete (see Fig. 1.17).
2. The restoration action implies debugging activity, which is performed perfectly with probability $a(0 < a \leq 1)$ and imperfectly with probability $b(= 1 - a)$. We call a the perfect debugging rate. One fault is corrected and removed from the software system when the debugging activity is perfect.
3. When n faults have been corrected, the time to the next software failure-occurrence and the restoration time follow exponential distributions with means of $1/\lambda_n$ and $1/\mu_n$, respectively.
4. The probability that two or more software failures will occur simultaneously is negligible.

Consider a stochastic process $\{X(t), t \geq 0\}$ with the state space (W, R) where up state vector $W = \{W_n; n = 0, 1, 2, \ldots\}$ and down state vector $R = \{R_n; n$

$= 0, 1, 2, \ldots\}$. Then, the events $\{X(t) = W_n\}$ and $\{X(t) = R_n\}$ mean that the system is operating and inoperable, respectively, due to the restoration action at time t, when n faults have already been corrected.

From assumption 2, when the restoration action has been completed in $\{X(t) = R_n\}$,

$$X(t) = \begin{cases} W_n, & \text{with probability } b, \\ W_{n+1}, & \text{with probability } a. \end{cases} \tag{1.30}$$

We use the Moranda model discussed in Sect. 1.3 to describe the software failure-occurrence phenomenon, i.e. when n faults have been corrected, the software hazard rate λ_n (see Fig. 1.14) is given by

$$\lambda_n = Dk^n, \quad n = 0, 1, 2, \ldots; D > 0, 0 < k < 1. \tag{1.31}$$

The expression of (1.31) comes from the point of view that software reliability depends on the debugging efforts, not the residual fault content. We do not note how many faults remain in the software system.

Next, we describe the time-dependent behavior of the restoration action. The restoration action for software systems includes not only the data recovery and the program reload, but also the debugging activities for manifested faults. From the viewpoint of the complexity, there are cases where the faults detected during the early stage of the testing or operation phase have low complexity and are easy to correct/remove, and as the testing is in progress, detected faults have higher complexity and are more difficult to correct/remove [8]. In the above case, it is appropriate that the mean restoration time becomes longer with the increasing number of corrected

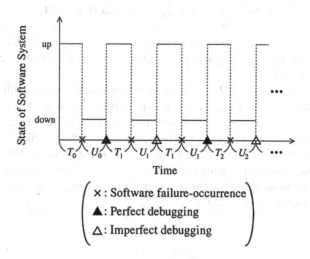

Fig. 1.17 Sample behavior of the software system alternating between up and down states

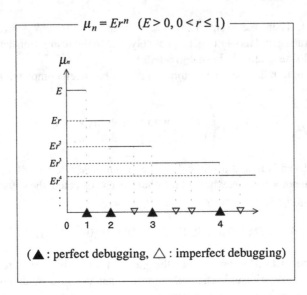

$$\mu_n = Er^n \quad (E > 0, 0 < r \le 1)$$

($\blacktriangle$: perfect debugging, $\triangle$: imperfect debugging)

Fault complexity
early-detected : easier to correct.
later-detected : more difficult to correct.

Fig. 1.18 Behavior of restoration rate

faults. Accordingly, we express μ_n as follows (see Fig. 1.18):

$$\mu_n = Er^n, \quad n = 0, 1, 2, \ldots; E > 0, 0 < r \le 1, \tag{1.32}$$

where E and r are the initial restoration rate and the decreasing ratio of the restoration rate, respectively. In (1.32) the case of $r = 1$, i.e. $\mu_n = E$, means that the complexity of each fault is random.

Let T_n and $U_n(n = 0, 1, 2, \ldots)$ be the random variables representing the next software failure-occurrence and the next restoration time-intervals when n faults have been corrected, in other words the sojourn times in states W_n and R_n, respectively. Furthermore, let $Y(t)$ be the random variable representing the cumulative number of faults corrected up to time t. The sample behavior of $Y(t)$ is illustrated in Fig. 1.19. It is noted that the cumulative number of corrected faults is not always coincident with that of software failures or restoration actions. The sample state transition diagram of $X(t)$ is illustrated in Fig. 1.20.

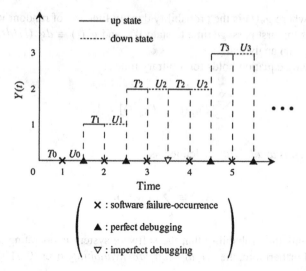

Fig. 1.19 A sample realization of $Y(t)$

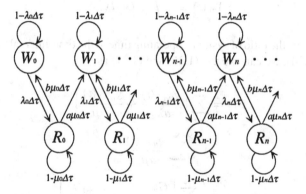

Fig. 1.20 A state transition diagram for software availability modeling

1.5.2 Software Availability Measures

We can obtain the state occupancy probabilities that the system is in states W_n and R_n at time point t as

$$P_{W_n}(t) \equiv \Pr\{X(t) = W_n\}$$
$$= \frac{g_{n+1}(t)}{a\lambda_n} + \frac{g'_{n+1}(t)}{a\lambda_n \mu_n}, \quad n = 0, 1, 2, \ldots, \tag{1.33}$$
$$P_{R_n}(t) \equiv \Pr\{X(t) = R_n\}$$
$$= \frac{g_{n+1}(t)}{a\mu_n}, \quad n = 0, 1, 2, \ldots, \tag{1.34}$$

respectively, where $g_n(t)$ is the probability density function of random variable S_n, which denotes the first passage time to state W_n, and $g'_n(t) \equiv dg_n(t)/dt$. $g_n(t)$ and $g'_n(t)$ can be given analytically.

The following equation holds for arbitrary time t:

$$\sum_{n=0}^{\infty} [P_{W_n}(t) + P_{R_n}(t)] = 1. \tag{1.35}$$

The *instantaneous availability* is defined as

$$A(t) \equiv \sum_{n=0}^{\infty} P_{W_n}(t), \tag{1.36}$$

which represents the probability that the software system is operating at specified time point t. Furthermore, the *average software availability* over $(0, t]$ is defined as

$$A_{av}(t) \equiv \frac{1}{t} \int_0^t A(x)dx, \tag{1.37}$$

which represents the ratio of system's operating time to the time-interval $(0, t]$. Using (1.33) and (1.34), we can express (1.36) and (1.37) as

$$A(t) = \sum_{n=0}^{\infty} \left[\frac{g_{n+1}(t)}{a\lambda_n} + \frac{g'_{n+1}(t)}{a\lambda_n \mu_n} \right]$$

$$= 1 - \sum_{n=0}^{\infty} \frac{g_{n+1}(t)}{a\mu_n}, \tag{1.38}$$

$$A_{av}(t) = \frac{1}{t} \sum_{n=0}^{\infty} \left[\frac{G_{n+1}(t)}{a\lambda_n} + \frac{g_{n+1}(t)}{a\lambda_n \mu_n} \right]$$

$$= 1 - \frac{1}{t} \sum_{n=0}^{\infty} \frac{G_{n+1}(t)}{a\mu_n}, \tag{1.39}$$

respectively, where $G_n(t)$ is the distribution function of S_n.

Figures 1.21 and 1.22 show numerical illustrations of $A(t)$ and $A_{av}(t)$ in (1.38) and (1.39), respectively.

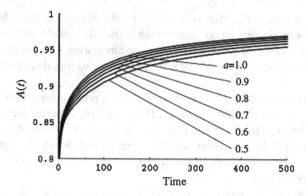

Fig. 1.21 Dependence of perfect debugging rate a on $A(t)$

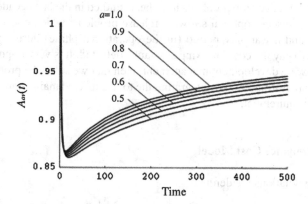

Fig. 1.22 Dependence of perfect debugging rate a on $A_{av}(t)$

1.6 Application of Software Reliability Assessment

It is very important to apply the results of software reliability assessment to management problems with software projects for attaining higher productivity and quality. We discuss three software management problems as application technologies of software reliability models.

1.6.1 Optimal Software Release Problem

Recently, it is becoming increasingly difficult for the developers to produce highly reliable software systems efficiently. Thus, it has been necessary to control a software development process in terms of quality, cost, and release time. In the last phase of the software development process, testing is carried out to detect and fix software faults introduced by human work, prior to its release for the operational use. The software faults that cannot be detected and fixed remain in the released software system after the testing phase. Thus, if a software failure occurs during the operational phase, then a computer system stops working and it may cause serious damage in our daily life.

If the duration of software testing is long, we can remove many software faults in the system and its reliability increases. However, this increases the testing cost and delays software delivery. In contrast, if the length of software testing is short, a software system with low reliability is delivered and it includes many software faults which have not been removed in the testing phase. Thus, the maintenance cost during the operation phase increases.

It is therefore very important in terms of software management that we find the optimal length of software testing, which is called an *optimal software release problem* [50–57]. These decision problems have been studied in the last decade by many researchers. We discuss optimal software release problems which consider both a present value and a warranty period (in the operational phase) during which the developer has to pay the cost for fixing any faults detected. It is very important with respect to software development management, then that we solve the problem of an optimal software testing time by integrating the total expected maintenance cost and the reliability requirement.

1.6.1.1 Maintenance Cost Model

The following notations are defined:

c_0 the cost for the minimum quantity of testing which must be done,
c_t the testing cost per unit time,
c_w the maintenance cost per one fault during the warranty period,
T the software release time, i.e. additional total testing time,
T^* the optimum software release time.

We discuss a *maintenance cost model* for formulation of the optimal release problem. The maintenance cost during the warranty period is considered. The concept of a present value is also introduced into the cost factors. Then, the total expected software maintenance cost $WC(T)$ can be formulated as:

$$WC(T) \equiv c_0 + c_t \int_0^T e^{-\alpha t} dt + C_w(T), \qquad (1.40)$$

where $C_w(T)$ is the maintenance cost during the warranty period. The parameter α in (1.40) is a discount rate of the cost. When we apply an exponential software reliability growth model based on an NHPP with mean value function $m(t)$ and intensity function $h_m(t)$ discussed in Sect. 1.3 (see Table 1.1), we discuss the following three cases in terms of the behavior of $C_w(T)$ (see Fig. 1.23):

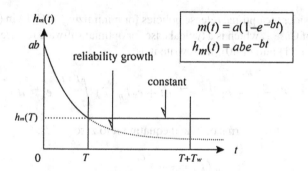

Fig. 1.23 Software reliability growth aspects during the warranty period

(Case 1)
When the length of the warranty period is constant and the software reliability growth is not assumed to occur after the testing phase, $C_w(T)$ is represented as:

$$C_w(T) = c_w \int_T^{T+T_w} h_m(T)e^{-\alpha t} dt. \tag{1.41}$$

(Case 2)
When the length of the warranty period is constant and the software reliability growth is assumed to occur even after testing, $C_w(T)$ is given by:

$$C_w(T) = c_w \int_T^{T+T_w} h_m(t)e^{-\alpha t} dt. \tag{1.42}$$

(Case 3)
When the length of the warranty period obeys a distribution function $W(t)$ and the software reliability growth is assumed to occur even after the testing phase, $C_w(T)$ is represented as:

$$C_w(T) = c_w \int_0^\infty \int_T^{T+T_w} h_m(t)e^{-\alpha t} dt\, dW(T_w), \tag{1.43}$$

where we assume that the distribution of the warranty period is a truncated normal distribution:

$$\frac{dW(t)}{dt} = \frac{1}{A\sqrt{2\pi}\sigma} \exp[-(t-\mu)^2/(2\sigma^2)], \quad t \geq 0, \mu > 0, \sigma > 0, \tag{1.44}$$

$$A = \frac{1}{\sqrt{2\pi}\sigma} \int_0^\infty \exp[-(t-\mu)^2/(2\sigma^2)] dt. \tag{1.45}$$

Let us consider the optimal release policies for minimizing $WC(T)$ in (1.40) with respect to T of Case 1, which is a typical case for optimal software release problems. Substituting (1.41) into (1.40), we rewrite it as:

$$WC(T) = c_0 + c_1 \int_0^T e^{-\alpha t} dt + c_w h_m(T) \int_T^{T+T_w} e^{-\alpha t} dt. \qquad (1.46)$$

Differentiating (1.46) in terms of T and equating it to zero yields:

$$h_m(T) = \frac{c_t}{c_w T_w (b+a)}. \qquad (1.47)$$

Note that $WC(T)$ is a convex function with respect to T because $d^2 WC(T)/dT^2 > 0$. Thus, the equation $dWC(T)/dT = 0$ has only one finite solution when the condition $h_m(0) > c_t/[c_w T_w(b+\alpha)]$ holds. The solution T_1 of (1.47) and the optimum release time can be shown as follows:

$$T^* = T_1 = \frac{1}{b} \ln \left[\frac{abc_w T_w(b+\alpha)}{c_t} \right], \quad 0 < T_1 < \infty. \qquad (1.48)$$

When the condition $h_m(0) \le c_t/[c_w T_w(b+\alpha)]$ holds, $WC(T)$ in (1.46) is a monotonically increasing function in terms of the testing time T. Then, the optimum release time $T^* = 0$. Therefore, we can obtain the optimal release policies as follows:

[Optimal Release Policy 1]

(1.1) If $h_m(0) > c_t/[c_w T_w(b+\alpha)]$, then the optimum release time is $T^* = T_1$.
(1.2) If $h_m(0) \le c_t/[c_w T_w(b+\alpha)]$, then the optimum release time is $T^* = 0$.

Similarly, we can obtain the optimal release policies for Case 2 and Case 3 [55, 58].

1.6.1.2 Maintenance Cost Model with Reliability Requirement

Next, we discuss the optimal release problem with the requirement for software reliability. In the actual software development, the manager must spend and control the testing resources with a view to minimizing the total software cost and satisfying reliability requirements rather than only minimizing the cost. From the exponential software reliability growth model, the software reliability function can be defined as the probability that a software failure does not occur during the time interval $(T, T+x]$ after the total testing time T, i.e. the release time. The software reliability function is given as follows:

$$R(x|T) = \exp[-\{m(T+x) - m(T)\}]. \qquad (1.49)$$

From (1.49), we derive the software reliability function as follows:

$$R(x|T) = \exp[-e^{-bT} \cdot m(x)].$$ (1.50)

Let the software reliability objective be $R_0(0 < R_0 \leq 1)$. We can evaluate optimum release time $T = T^*$ which minimizes (1.40) while satisfying the software reliability objective R_0. Thus, the optimal software release problem is formulated as follows:

$$\text{minimize } WC(T) \quad \text{subject to} \quad R(x|T) \geq R_0.$$ (1.51)

For the optimal release problem formulated by (1.51), let T_R be the optimum release time with respect to T satisfying the relation $R(x|T) = R_0$ for specified x. By applying the relation $R(x|T) = R_0$ into (1.50), we can obtain the solution T_R as follows:

$$T_R = \frac{1}{b}\left\{\ln m(x) - \ln\ln\left(\frac{1}{R_0}\right)\right\}.$$ (1.52)

Then, we can derive the optimal release policies to minimize the total expected software maintenance cost and to satisfy the software reliability objective R_0.

For Case 1, the optimal release policies are given as follows:

[Optimal Release Policy 2]

(2.1) If $h_m(0) > c_t/[c_w T_w(b + \alpha)]$ and $R(x|0) < R_0$, then the optimum release time is $T^* = \max\{T_1, T_R\}$.

(2.2) If $h_m(0) > c_t/[c_w T_w(b + \alpha)]$ and $R(x|0) \geq R_0$, then the optimum release time is $T^* = T_1$.

(2.3) If $h_m(0) \leq c_t/[c_w T_w(b + \alpha)]$ and $R(x|0) < R_0$, then the optimum release time is $T^* = T_R$.

(2.4) If $h_m(0) \leq c_t/[c_w T_w(b + \alpha)]$ and $R(x|0) \geq R_0$, then the optimum release time is $T^* = 0$.

Similarly, we can obtain the optimal release policies for Case 2 and Case 3 [58]. Figure 1.24 shows numerical illustrations of the optimum release time in [Optimal Release Policy 2], where $T^* = \max\{T_1, T_R\} = \max\{92.5, 122.8\} = 122.8$.

1.6.2 Statistical Software Testing-Progress Control

As well as quality/reliability assessment, software-testing managers should assess the degree of testing-progress. We can construct a statistical method for software testing-progress control based on a *control chart* method as follows [6, 59]. This method is based on several instantaneous fault-detection rates derived from software

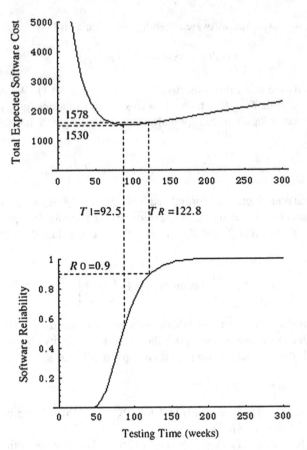

Fig. 1.24 Optimal software release time for optimal release policy 2 ($c_0 = 1000.0$, $c_t = 5.0$, $c_w = 20.0$, $a = 1000$, $b = 0.05$, $\mu = 100$, $\omega = 10$, $\alpha = 0.001$, $x = 1.0$, $R_0 = 0.9$)

reliability growth models based on an NHPP. For example, the intensity function based on the delayed S-shaped software reliability growth model in Sect. 1.3 (see Table 1.1) is given by

$$h_M(t) = \frac{dM(t)}{dt} = ab^2te^{-bt}, \quad a > 0, b > 0. \tag{1.53}$$

From (1.53), we can derive

$$\ln Z_M(t) = \ln a + 2 \cdot \ln b - bt, \tag{1.54}$$

$$Z_M(t) = \frac{h_M(t)}{t}. \tag{1.55}$$

The mean value of the instantaneous fault-detection rate represented by (1.55) is defined as the *average-instantaneous fault-detection rate*. Equation (1.54) means that the relation between the logarithm value of $Z_M(t)$ and the testing time has a linear property. If the testing phase progresses smoothly and the reliability growth is stable in the testing, the logarithm of the average-instantaneous fault-detection rate decreases linearly with the testing time. From (1.54), we can also estimate the unknown parameters a and b by the method of least-squares, and assess the testing progress by applying a regression analysis to the observed data. It is assumed that the form of the data is $(t_k, Z_k)(k = 1, 2, \ldots, n)$ where t_k is the kth testing time and Z_k is the realization of average-instantaneous fault-detection rate $Z_M(t)$ at testing-time t_k. Letting the estimated unknown parameters be $\hat{a}$ and $\hat{b}$, we obtain the estimator of $Y(= \ln Z_M(t))$ as follows:

$$\hat{Y} = \ln \hat{Z}_M(t) = \ln \hat{a} + 2 \cdot \ln \hat{b} - \hat{b}t = \bar{Y} - \hat{b}(t - \bar{t}), \tag{1.56}$$

where

$$\bar{Y} = \frac{1}{n} \sum_{k=1}^{n} Y_k, \quad Y_k = \ln Z_k, \quad \bar{t} = \frac{1}{n} \sum_{k=1}^{n} t_k, \quad k = 1, 2, \ldots, n.$$

The variation, which is explained as the regression to the dependent variable, Y, is

$$S_b = \sum_{k=1}^{n} (\hat{Y}_k - \bar{Y})^2 = \hat{b}^2 \sum_{k=1}^{n} (t_k - \bar{t})^2. \tag{1.57}$$

On the other hand, the error-variation not explained as the regression is represented as the summation of residual squares. That is,

$$S_e = \sum_{k=1}^{n} (Y_k - \hat{Y}_k)^2. \tag{1.58}$$

The unbiased variances from (1.57) and (1.58) are:

$$V_b = S_b, \quad V_e = \frac{S_e}{n-2}. \tag{1.59}$$

With reference to (1.56), we discuss the logarithm of average-instantaneous fault-detection rate $Y_0 = \ln Z_M(t_0)$ at $t = t_0(t_0 \geq t_n)$ by using the results of the analysis of variance. The $100(1 - \alpha)$ percent confidence interval to $\hat{Y}_0$ is given by

$$\hat{Y}_0 \pm t \left(n - 2, 1 - \frac{\alpha}{2} \right) \sqrt{\text{Var}[\hat{Y}_0]},$$

$$\text{Var}[\hat{Y}_0] = \left\{ 1 + \frac{1}{n} + \frac{(t_0 - \bar{t})^2}{\sum_{k=1}^{n}(t_k - \bar{t})^2} \right\} V_e. \tag{1.60}$$

$\text{Var}[\hat{Y}_0]$ in (1.60) is the variance of $\hat{Y}_0$. $t(h, p)$ in (1.60) is $100p$ percent point of t-distribution at degree of freedom h. We now make the control chart which consists of the center line by the logarithm of the average-instantaneous fault-detection rate, and the upper and lower control limits which are given by (1.60). We can assess the testing-progress by applying a regression analysis to the observed data.

The testing-progress assessment indices for the other NHPP models are given by the following intensity function:

- $h_m(t) = abe^{-bt}$ with relation $\ln h_m(t) = (\ln a + \ln b) - bt$ (for the exponential software reliability growth model),
- $h_\mu(t) = \lambda_0/(\lambda_0 \theta t + 1)$ with relation $\ln h_\mu(t) = \ln \lambda_0 - \theta \mu(t)$ (for the logarithmic Poisson execution time model),
- $h_\lambda(t) = \lambda \beta t^{\beta-1}$ with relation $\ln h_\lambda(t) = (\ln \lambda + \ln \beta) + (\beta - 1) \ln t$ (for the Weibull process model [6, 8]).

The procedure of testing-progress control is shown as follows:

Step 1: An appropriate model is selected to apply and the model parameters are estimated by the method of least-squares.

Step 2: To certify goodness-of-fit of the estimated regression equation for the observed data, we use the F-test.

Step 3: Based on the result of the F-test, the central line and upper and lower control limits of the control chart are calculated. The control chart is drawn.

Step 4: The observed data are plotted on the control chart and the stability of the testing-progress is judged.

Figure 1.25 shows examples of the control chart for the testing-progress control based on the delayed S-shaped software reliability growth model.

1.6.3 Optimal Testing-Effort Allocation Problem

We discuss a management problem to achieve a reliable software system efficiently during module testing in the software development process by applying a testing-effort-dependent software reliability growth model based on an NHPP (see Table 1.1). We take account of the relationship between the testing-effort spent during the module testing and the detected software faults where the testing-effort is defined as resource expenditures spent on software testing, e.g. manpower, CPU hours, and executed test cases. The software development manager has to decide how to use the specified testing-effort effectively in order to maximize the software quality and reliability [60].

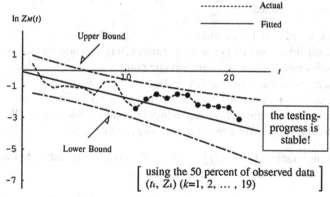

[An example of the control chart for the testing-progress control]

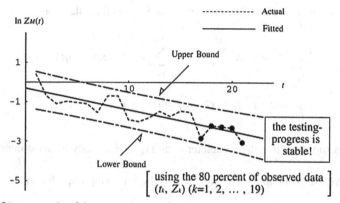

[An example of the control chart for the testing-progress control]

Fig. 1.25 Examples of the control chart for the testing-progress control

That is, to develop a quality and reliable software system, it is very important for the manager to allocate the specified amount of testing-effort expenditure for each software module under some constraints. We can observe the software reliability growth in the module testing in terms of a time-dependent behavior of the cumulative number of faults detected during the testing stage.

Based on the testing-effort dependent software reliability growth model, we consider the following testing-effort allocation problem [61, 62]:

1. The software system is composed of M independent modules. The number of software faults remaining in each module can be estimated by the model.
2. The total amount of testing-effort expenditure for module testing is specified.
3. The manager has to allocate the specified total testing-effort expenditure to each software module so that the number of software faults remaining in the system may be minimized.

The following are defined:

a the expected initial fault content,
r the fault-detection rate per unit of testing-effort expenditure ($0 < r < 1$),
i the subscript for each software module number $i = 1, 2, \ldots, M$,
w_i the weight for each module ($w_i > 0$),
n_i the expected number of faults remaining in each module,
q_i, Q the amount of testing-effort expenditure for each module to be allocated and the total testing-effort expenditure before module testing ($q_i \geq 0, Q > 0$).

From (1.8) and Table 1.1, i.e. $n(t) = a \cdot \exp[-rW(t)]$, the estimated number of remaining faults for module i is formulated by

$$n_i = a_i \cdot \exp[-r_i q_i], \quad i = 1, 2, \ldots M. \tag{1.61}$$

Thus, the *optimal testing-effort allocation problem* is formulated as:

$$\text{minimize} \sum_{i=1}^{M} w_i n_i = \sum_{i=1}^{M} w_i a_i \cdot \exp[-r_i q_i], \tag{1.62}$$

$$\text{so that} \sum_{i=1}^{M} q_i \leq Q, \quad q_i \geq 0, \quad i = 1, 2, \ldots, M, \tag{1.63}$$

where it is supposed that the parameter a_i and r_i have already been estimated by the model.

To solve the problem above, we consider the following Lagrangian:

$$L = \sum_{i=1}^{M} w_i a_i \cdot \exp[-r_i q_i] + \lambda \left(\sum_{i=1}^{M} q_i - Q \right), \tag{1.64}$$

and the necessary and sufficient conditions [63] for the minimum are

$$\left. \begin{aligned} \frac{\partial L}{\partial q_i} &= -w_i a_i r_i \cdot \exp[-r_i q_i] + \lambda \geq 0, \\ q_i \cdot \frac{\partial L}{\partial q_i} &= 0, \quad i = 1, 2, \ldots, M, \\ \sum_{i=1}^{M} q_i &= Q, \\ q_i &\geq 0, \quad i = 1, 2, \ldots, M, \end{aligned} \right\} , \tag{1.65}$$

where λ is a Lagrange multiplier.

Without loss of generality, setting $A_i = w_i a_i r_i (i = 1, 2, \ldots, M)$, we can assume that the following condition is satisfied for the tested modules:

$$A_1 \geq A_2 \geq \cdots \geq A_{k-1} \geq A_k \geq A_{k+1} \geq \cdots \geq A_M. \tag{1.66}$$

This means that it is arranged in order of fault detectability for the tested modules. Now, if $A_k > \lambda \geq A_{k+1}$, from (1.65) we have

$$q_i = \max\left\{0, \frac{1}{r_i}(\ln A_i - \ln \lambda)\right\},$$

i.e.

$$\left.\begin{array}{ll} q_i = \dfrac{1}{r_i}(\ln A_i - \ln \lambda), & i = 1, 2, \ldots, k, \\ q_i = 0, & i = k+1, \ldots, M, \end{array}\right\} \tag{1.67}$$

From (1.65) and (1.67), $\ln \lambda$ is given by

$$\ln \lambda = \frac{\sum_{i=1}^{k} \frac{1}{r_i} \ln A_i - Q}{\sum_{i=1}^{k} \frac{1}{r_i}}, \quad k = 1, 2, \ldots, M. \tag{1.68}$$

Let λ_k denote the value of the right-hand side of (1.68). Then, the optimal Lagrange multiplier λ^* exists in the set $\{\lambda_1, \lambda_2, \ldots, \lambda_M\}$. Hence, we can obtain λ^* by the following procedures:

(1) Set $k = 1$.
(2) Compute λ_k by (1.68).
(3) If $A_k > \lambda_k \geq A_{k+1}$, then $\lambda^* = \lambda_k$ (stop). Otherwise, set $k = k + 1$ and go back to (2).

The optimal solutions $q_i^* (i = 1, 2, \ldots, M)$ are given by

Table 1.2 Example of the optimal testing-effort allocation problem ($M = 10$, $Q = 5300$)

Software module i	a_i	w_i	r_i	q_i^*	z_i
1	1000	1.0	$1.0 (\times 10^{-2})$	191.7	147.0
2	1000	1.0	1.0	191.7	147.0
3	1000	1.0	1.0	191.7	147.0
4	500	1.0	0.5	106.2	294.0
5	500	1.0	0.5	106.2	294.0
6	500	1.0	0.5	106.2	294.0
7	500	1.0	0.5	106.2	294.0
8	100	1.0	0.1	0.0	100.0
9	100	1.0	0.1	0.0	100.0
10	100	1.0	0.1	0.0	100.0
Total	5300				1917.0

$$\left. \begin{array}{ll} q_i^* = \frac{1}{r_i}(\ln A_i - \ln \lambda^*), & i = 1, 2, \ldots, k, \\ q_i^* = 0, & i = k + 1, \ldots, M, \end{array} \right\}, \qquad (1.69)$$

which means that the amount of testing-effort expenditure is needed more for the tested modules containing more faults.

Table 1.2 shows numerical examples of the optimal testing-effort allocation problem.

References

1. Kanno, A. (1992). *Introduction to software production engineering (in Japanese)*. Tokyo: JUSE Press.
2. Matsumoto, Y., & Ohno, Y. (Eds.). (1989). *Japanese perspectives in software engineering*. Singapore: Addison-Wesley.
3. Lyu, M. R. (Ed.). (1996). *Handbook of software reliability engineering*. Los Alamitos, CA: IEEE Computer Society Press.
4. Pham, H. (2006). *System software reliability*. London: Springer.
5. Yamada, S., & Ohtera, H. (1990). *Software reliability: Theory and practical application (in Japanese)*. Tokyo: Soft Research Center.
6. Yamada, S. (1994). *Software reliability models: Fundamentals and applications (in Japanese)*. Tokyo: JUSE Press.
7. Yamada, S. (2011). *Elements of software reliability—modeling approach (in Japanese)*. Tokyo: Kyoritsu-Shuppan.
8. Musa, J. D., Iannino, A., & Okumoto, A. (1987). *Software reliability: Measurement, prediction, application*. New York: McGraw-Hill.
9. Ramamoorthy, C. V., & Bastani, F. B. (1982). Software reliability–status and perspectives. *IEEE Transactions on Software Engineering, SE-8*, 354–371.
10. Jelinski, Z., & Moranda, P. B. (1972). Software reliability research. In W. Freiberger (Ed.), *Statistical computer performance evaluation* (pp. 465–484). New York: Academic Press.
11. Wagoner, W. L. (1973). The final report on a software reliability measurement study. *Report TOR-0074(4112)-1*, Aerospace Corporation.
12. Moranda, P. B. (1979). Event-altered rate models for general reliability analysis. *IEEE Transactions on Reliability, R-28*, 376–381.
13. Ascher, H., & Feingold, H. (1984). *Repairable systems reliability: Modeling, inference, misconceptions, and their causes*. New York: Marcel Dekker.
14. Yamada, S. (1991). Software quality/reliability measurement and assessment: software reliability growth models and data analysis. *Journal of Information Processing, 14*, 254–266.
15. Yamada, S., & Osaki, S. (1985). Software reliability growth modeling: Models and applications. *IEEE Transactions on Software Engineering, SE-11*, 1431–1437.
16. Goel, A. L., & Okumoto, K. (1979). Time-dependent error-detection rate model for software reliability and other performance measures. *IEEE Transactions on Reliability, R-28*, 206–211.
17. Goel, A. L. (1980). Software error detection model with applications. *Journal of Systems and Software, 1*, 243–249.
18. Yamada, S., & Osaki, S. (1984). Nonhomogeneous error detection rate models for software reliability growth. In S. Osaki & Y. Hatoyama (Eds.), *Stochastic models in reliability theory* (pp. 120–143). Berlin: Springer.
19. Yamada, S., Osaki, S., & Narihisa, H. (1985). A software reliability growth model with two types of errors. *R. A. I. R. O. Operations Research, 19*, 87–104.
20. Yamada, S., Ohba, M., & Osaki, S. (1983). S-shaped reliability growth modeling for software error detection. *IEEE Transactions on Reliability, R-32*, 475–478, 484.

21. Yamada, S., Ohba, M., & Osaki, S. (1984). S-shaped software reliability growth models and their applications. *IEEE Transactions on Reliability, R-33*, 289–292.
22. Ohba, M. (1984). Inflection S-shaped software reliability growth model. In S. Osaki & Y. Hatoyama (Eds.), *Stochastic models in reliability theory* (pp. 144–162). Berlin: Springer.
23. Ohba, M., & Yamada, S. (1984). S-shaped software reliability growth models. In *Proceedings of the 4th International Conference on Reliability and Maintainability* (pp. 430–436).
24. Yamada, S., Ohtera, H., & Narihisa, H. (1986). Software reliability growth models with testing-effort. *IEEE Transactions on Reliability, R-35*, 19–23.
25. Yamada, S., Hishitani, J., & Osaki, S. (1993). Software-reliability growth with a Weibull test-effort function. *IEEE Transactions on Reliability, R-42*, 100–106.
26. Ohtera, H., Yamada, S., & Ohba, M. (1990). Software reliability growth model with testing-domain and comparisons of goodness-of-fit. In *Proceedings of the International Symposium on Reliability and Maintainability* (pp. 289–294).
27. Yamada, S., Ohtera, H., & Ohba, M. (1992). Testing-domain dependent software reliability growth models. *Computers & Mathematics with Applications, 24*, 79–86.
28. Musa, J. D., & Okumoto, K. (1984). A logarithmic Poisson execution time model for software reliability measurement. In *Proceedings of the 7th International Conference on Software Engineering* (pp. 230–238).
29. Okumoto, K. (1985). A statistical method for software quality control. *IEEE Transactions on Software Engineering, SE-11*, 1424–1430.
30. Kanno, A. (1979). *Software engineering (in Japanese)*. Tokyo: JUSE Press.
31. Mitsuhashi, T. (1981). *A method of software quality evaluation (in Japanese)*. Tokyo: JUSE Press.
32. Shoomam, M. L. (1983). *Software engineering: Design, reliability, and management*. New York: McGraw-Hill.
33. Ohba, M., & Chou, X. (1989). Does imperfect debugging affect software reliability growth? In *Proceedings of the 11th International Conference on Software Engineering* (pp. 237–244).
34. Shanthikumar, J. G. (1981). A state- and time-dependent error occurrence-rate software reliability model with imperfect debugging. In *Proceedings of the National Computer Conference* (pp. 311–315).
35. Ross, S. M. (1996). *Stochastic processes*. New York: Wiley.
36. Yamada, S., & Miki, T. (1998). Imperfect debugging models with introduced software faults and their comparisons (in Japanese). *Transactions of IPS Japan, 39*, 102–110.
37. Yamada, S. (1998). Software reliability growth models incorporating imperfect debugging with introduced faults. *Electronics and Communications in Japan, 81*, 33–41.
38. Xie, M. (1991). *Software reliability modeling*. Singapore: World Scientific.
39. Laprie, J.-C., Kanoun, K., Béounes, C., & Kaâniche, M. (1991). The KAT (Knowledge-Action-Transformation) approach to the modeling and evaluation of reliability and availability growth. *IEEE Transactions on Software Engineering, 17*, 370–382.
40. Laprie, J.-C., & Kanoun, K. (1992). X-ware reliability and availability modeling. *IEEE Transactions on Software Engineering, 18*, 130–147.
41. Tokuno, K., & Yamada, S. (1995). A Markovian software availability measurement with a geometrically decreasing failure-occurrence rate. *IEICE Transactions on Fundamentals of Electronics, Communications and Computer Sciences, E78-A*, 737–741.
42. Okumoto, K., & Goel, A. L. (1978). Availability and other performance measures for system under imperfect maintenance. In *Proceedings of the COMPSAC '78* (pp. 66–71).
43. Kim, J. H., Kim, Y. H., & Park, C. J. (1982). A modified Markov model for the estimation of computer software performance. *Operations Research Letters, 1*, 253–257.
44. Tokuno, K., & Yamada, S. (1997). Markovian software availability modeling for performance evaluation. In A. H. Christer, S. Osaki, & L. C. Thomas (Eds.), *Stochastic modeling in innovative manufacturing* (pp. 246–256). Berlin: Springer.
45. Tokuno, K., & Yamada, S. (2007). User-oriented and -perceived software availability measurement and assessment with environment factors. *Journal of Operations Research Society of Japan, 50*, 444–462.

46. Tokuno, K., & Yamada, S. (2011). Codesign-oriented performability modeling for hardware-software systems. *IEEE Transactions on Reliability, 60*, 171–179.
47. Tokuno, K., & Yamada, S. (1997). Markovian software availability modeling with two types of software failures for operational use. In *Proceedings of the 3rd ISSAT International Conference on Reliability and Quality in Design* (pp. 97–101).
48. Tokuno, K., & Yamada, S. (2001). Markovian modeling for software availability analysis under intermittent use. *International Journal of Reliability, Quality and Safety Engineering, 8*, 249–258.
49. Tokuno, K., & Yamada, S. (1998). Operational software availability measurement with two kinds of restoration actions. *Journal of Quality in Maintenance Engineering, 4*, 273–283.
50. Cho, B. C., & Park, K. S. (1994). An optimal time for software testing under the user's requirement of failure-free demonstration before release. *IEICE Transactions on Fundamentals of Electronics, Communications, and Computer Sciences, E77-A*, 563–570.
51. Foreman, E. H., & Singpurwalla, N. D. (1979). Optimal time intervals for testing-hypotheses on computer software errors. *IEEE Transactions on Reliability, R-28*, 250–253.
52. Kimura, M., & Yamada, S. (1995). Optimal software release policies with random life-cycle and delivery delay. In *Proceedings of the 2nd ISSAT International Conference on Reliability and Quality in Design* (pp. 215–219).
53. Koch, H. S., & Kubat, P. (1983). Optimal release time for computer software. *IEEE Transactions on Software Engineering, SE-9*, 323–327.
54. Okumoto, K., & Goel, A. L. (1980). Optimum release time for software system based on reliability and cost criteria. *Journal of Systems and Software, 1*, 315–318.
55. Yamada, S. (1994). Optimal release problems with warranty period based on a software maintenance cost model (in Japanese). *Transactions of IPS Japan, 35*, 2197–2202.
56. Yamada, S., Kimura, M., Teraue, E., & Osaki, S. (1993). Optimal software release problem with life-cycle distribution and discount rate (in Japanese). *Transactions of IPS Japan, 34*, 1188–1197.
57. Yamada, S., & Osaki, S. (1987). Optimal software release policies with simultaneous cost and reliability requirements. *European Journal of Operational Research, 31*, 46–51.
58. Kimura, M., Toyota, T., & Yamada, S. (1999). Economic analysis of software release problems with warranty cost and reliability requirement. *Reliability Engineering and System Safety, 66*, 49–55.
59. Yamada, S., & Kimura, M. (1999). Software reliability assessment tool based on object-oriented analysis and its application. *Annals of Software Engineering, 8*, 223–238.
60. Kubat, P., & Koch, H. S. (1983). Managing test procedures to achieve reliable software. *IEEE Transactions on Reliability, R-32*, 299–303.
61. Ohtera, H., & Yamada, S. (1990). Optimal allocation and control problem for software testing-resources. *IEEE Transactions on Reliability, R-39*, 171–176.
62. Yamada, S., Ichimori, T., & Nishiwaki, N. (1995). Optimal allocation policies for testing-resource based on a software reliability growth model. *Mathematical and Computer Modeling, 22*, 295–301.
63. Bazaraa, M. S., & Shetty, C. M. (1979). *Nonlinear programming: Theory and algorithms.* New York: Wiley.

Chapter 2
Recent Developments in Software Reliability Modeling

Abstract Management technologies for improving software reliability are very important for software TQM (Total Quality Management). The quality characteristics of software reliability is that computer systems can continue to operate regularly without the occurrence of failures on software systems. In this chapter, we describe several recent developments in software reliability modeling and its applications as quantitative techniques for software quality/reliability measurement and assessment. That is, a quality engineering analysis of human factors affecting software reliability during the design-review phase, which is the upper stream of software development, and software reliability growth models based on stochastic differential equations and discrete calculus during the testing phase, which is the lower one, are discussed. And, we discuss quality-oriented software management analysis by applying the multivariate analysis method and the existing software reliability growth models to actual process monitoring data. Finally, we investigate an operational performability evaluation model for the software-based system, introducing the concept of systemability which is defined as the reliability characteristic subject to the uncertainty of the field environment.

Keywords Human factor analysis · Design-review experiment · OSS reliability · Stochastic differential equation · Discrete modeling · Difference equation · Software management · Software project assessment · Software performability modeling · Systemability assessment

2.1 Introduction

At present, it is important to assess the reliability of software systems because of increasing the demands on quality and productivity in social systems. Moreover, they may cause serious accidents affecting people's lives. Under the background like this, software reliability technologies for the purpose of producing quality software systems efficiently, systematically, and economically have been developed and researched energetically. Especially, comprehensive use of technologies and methodologies in software engineering is needed for improving software quality/reliability.

A computer-software is developed by human work, therefore many software faults must be introduced into the software product during the development process. Thus, these software faults often cause break-downs of computer systems. Recently, it becomes more difficult for the developers to produce highly-reliable software systems efficiently because of the diversified and complicated software requirements. Therefore, it is necessary to control the software development process in terms of quality and reliability. Note that *software failure* is defined as an unacceptable departure of program operation caused by a *software fault* remaining in the software system.

First, we focus on a software design-review process which is more effective than the other processes in the upper stream of software development for elimination and prevention of software faults. Then conducting a design-review experiment, we discuss a quality engineering approach for analyzing the relationships among the quality of the design-review activities, i.e. software reliability, and human factors to clarify the fault-introduction process in the design-review process.

Basically, software reliability can be evaluated by the number of detected faults or the software failure-occurrence time in the testing phase which is the last phase of the development process, and it can be also estimated in the operational phase. Especially, *software reliability models* which describe software fault-detection or failure-occurrence phenomena in the system testing phase are called *software reliability growth models (SRGM's)* . The SRGM's are useful to assess the reliability for quality control and testing-progress control of software development. Most of the SRGM's which have been proposed up to the present treat the event of fault-detection in the testing and operational phases as a counting process. However, if the size of the software system is large, the number of faults detected during the testing phase become large, and the change of the number of faults which are detected and removed through debugging activities becomes sufficiently small compared with the initial fault content at the beginning of the testing phase.

Then, we model the fault-detection process as a stochastic process with a continuous state space for reliability assessment in an open source solution developed under several open source softwares (OSS's) to consider the active state of the open source projects and the collision among the open source components. And we propose a new SRGM describing the fault-detection process by applying a mathematical technique of stochastic differential equations of Itô-type.

Further, based on discrete analogs of nonhomogeneous Poisson process models as SRGM's, which have exact solutions in terms of solving the hypothesized differential equations, we propose two discrete models described by difference equations derived by transforming the continuous testing-time into discrete one. Then, we can show that such a difference calculus enables us to assess software reliability more accurately than conventional discrete models.

And, we discuss quality-oriented software management through statistical analysis of process monitoring data. Then, based on the desired software management models, we obtain the significant process factors affecting QCD (Quality, Cost, and

Delivery) measures. At the same time, we propose a method of software reliability assessment as process monitoring evaluation with actual data for the process monitoring progress ratio and the pointed-out problems (i.e. detected faults).

Finally, we investigate an operational performability evaluation model for the software-based system, introducing the concept of *systemability* which is defined as the reliability characteristic subject to the uncertainty of the field environment. Assuming that the software system can process the multiple tasks simultaneously and that the arrival process of the tasks follows a nonhomogeneous Poisson process, we analyze the distribution of the number of tasks whose processes can be completed within the processing time limit with the infinite-server queueing theory. Here we take the position that the software reliability characteristic in the testing phase is originally different from that in the operation phase. Then, the software failure-occurrence phenomenon in the operation phase is described with the Markovian software reliability model with systemability, i.e. we consider the randomness of the environmental factor which is introduced to bridge the gap between the software failure-occurrence characteristics during the testing and the operation phases. We derive several software performability measures considering the real-time property; these are given as the functions of time and the number of debugging activities. Finally, we illustrate several numerical examples of the measures to investigate the impact of consideration of systemability on the system performability evaluation.

2.2 Human Factors Analysis

In this section, we discuss an experiment study to clarify human factors [1–3] and their interactions affecting software reliability by assuming a model of human factors which consist of inhibitors and inducers. In this experiment, we focus on the software design-review process which is more effective than the other processes in the elimination and prevention of software faults. For an analysis of experimental results, a quality engineering approach base on a *signal-to-noise ratio* (defined as SNR) [4] is introduced to clarify the relationships among human factors and software reliability measured by the number of seeded faults detected by review activities, and the effectiveness of significant human factors judged by the design of experiment [5] is evaluated. As a result, applying the orthogonal array $L_{18}(2^1 \times 3^7)$ to the human factor experiment, we obtain the optimal levels for the selected inhibitors and inducers.

2.2.1 Design-Review and Human Factors

The inputs and outputs for the design-review process are shown in Fig. 2.1. The design-review process is located in the intermediate process between design and coding phases, and have software requirement-specifications as inputs and software

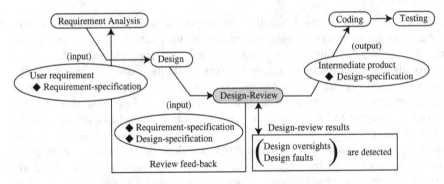

Fig. 2.1 Inputs and outputs in the software design process

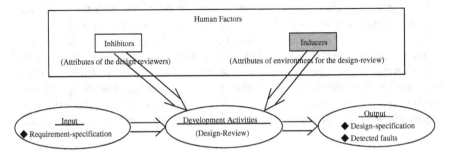

Fig. 2.2 A human factor model including inhibitors and inducers

design-specifications as outputs. In this process, software reliability is improved by detecting software faults effectively [6].

The attributes of software designers and design process environment are mutually related for the design-review process (see Fig. 2.1). Then, influential human factors for the design-specifications as outputs are classified into two kinds of attributes in the followings [7–9] (see Fig. 2.2):

(i) Attributes of the design reviewers (*Inhibitors*)

Attributes of the design reviewers are those of software engineers who are responsible for design-review work. For example, they are the degree of understanding of software requirement-specifications and software design-methods, the aptitude of programmers, the experience and capability of software design, the volition of achievement of software design, etc. Most of them are psychological human factors which are considered to contribute directly to the quality of software design-specification.

(ii) Attributes of environment for the design-review (*Inducers*)

In terms of design-review work, many kinds of influential factors are considered such as the education of software design-methods, the kind of software design methodologies, the physical environmental factors in software design work,

Table 2.1 Human factors in the design-review experiment

	Human factor	Level		
		1	2	3
$A^{(ii)}$	BGM of classical music the review work environment	A_1: yes	A_2: no	-
$B^{(ii)}$	Time duration of software design work (minute)	B_1: 20 min	B_2: 30 min	B_3: 40 min
$C^{(i)}$	Degree of understanding of the design method (R-Net Technique)	C_1: high	C_2: common	C_3: low
$D^{(i)}$	Degree of understanding of require-ment specifications	D_1: high	D_2: common	D_3: low
$E^{(ii)}$	Check list (indicating the matters that require attention in review work)	E_1: detailed	E_2: common	E_3: nothing

(i) Inhibitor, (ii) Inducers

e.g., temperature, humidity, noise, etc. All of these influential factors may affect indirectly to the quality of software design-specification.

2.2.2 Design-Review Experiment

In order to find out the relationships among the reliability of software design-specification and its influential human factors, we have performed the design of experiment by selecting five human factors as shown in Table 2.1.

In this experiment, we conduct an experiment to clarify the relationships among human factors affecting software reliability and the reliability of design-review work by assuming a human factor model consisting of inhibitors and inducers as shown in Fig. 2.2. The actual experiment has been performed by 18 subjects based on the same design-specification of a triangle program which receives three integers representing the sides of a triangle and classifies the kind of triangle such sides form [10]. We measured the 18 subjects' capability of both the degrees of understanding of design-method and requirement-specification by the preliminary tests before the design of experiment. Further, we seeded some faults in the design-specification intentionally. Then, we have executed such a design-review experiment in which the 18 subjects detect the seeded faults.

We have performed the experiment by using the five human factors with three levels as shown in Table 2.1, which are assigned to the orthogonal-array $L_{18}(2^1 \times 3^7)$ of the design of experiment as shown in Table 2.3. We distinguish the design parts as follows to be pointed out in the design-review as detected faults into the descriptive-design and symbolic-design parts.

- **Descriptive-design faults**
  The descriptive-design parts consist of words or technical terminologies which are described in the design-specification to realize the required functions. In this experiment, the descriptive-design faults are algorithmic ones, and we can improve the quality of design-specification by detecting and correcting them.

- **Symbolical-design faults**
  The symbolical-design parts consist of marks or symbols which are described in the design-specification. In this experiment, the symbolical-design faults are notation mistakes, and the quality of the design-specification can not be improved by detecting and correcting them.

For the orthogonal-array $L_{18}(2^1 \times 3^7)$ as shown in Table 2.3, setting the classification of detected faults as outside factor R and the human factors A, B, C, D, and E as inside factors, we perform the design-review experiment. Here, the outside factor R has two levels such as descriptive-design parts (R_1) and symbolical-design parts (R_2).

2.2.3 Analysis of Experimental Results

We define the efficiency of design-review, i.e. the reliability, as the degree that the design reviewers can accurately detect correct and incorrect design parts for the design-specification containing seeded faults. There exists the following relationship among the total number of design parts, n, the number of correct design parts, n_0, and the number of incorrect design parts containing seeded faults, n_1:

$$n = n_0 + n_1. \tag{2.1}$$

Therefore, the design parts are classified as shown in Table 2.2 by using the following notations:

Table 2.2 Input and output tables for two kinds of error

	(i)Observed values				(ii)Error rates		
Output Input	0 (true)	1 (false)	Total	Output Input	0 (true)	1 (false)	Total
0 (true)	n_{00}	n_{01}	n_0	0 (true)	$1 - p$	p	1
1 (false)	n_{10}	n_{11}	n_1	1 (false)	q	$1 - q$	1
Total	r_0	r_1	n	Total	$1 - p + q$	$1 - q + p$	2

Table 2.3 The orthogonal array $L_{18}(2^1 \times 3^7)$ with assigned human factors and experimental data

No.	A	B	C	D	E	R_1				R_2				SNR (dB) R_1	R_2
						n_{00}	n_{01}	n_{10}	n_{11}	n_{00}	n_{01}	n_{10}	n_{11}		
1	1	1	1	1	1	52	0	2	12	58	1	0	4	7.578	6.580
2	1	1	2	2	2	49	3	8	6	59	0	2	2	−3.502	3.478
3	1	1	3	3	3	50	2	12	2	59	0	4	0	8.769	2.342
4	1	2	1	1	2	52	0	2	12	59	0	0	4	7.578	8.237
5	1	2	2	2	3	50	2	4	10	57	2	0	4	1.784	4.841
6	1	2	3	3	1	45	7	8	6	59	0	3	1	−7.883	0.419
7	1	3	1	2	1	52	0	2	12	59	0	2	2	7.578	3.478
8	1	3	2	3	2	47	5	6	8	59	0	2	2	−3.413	3.478
9	1	3	3	1	3	52	0	10	4	58	1	1	3	0.583	4.497
10	2	1	1	3	3	52	0	10	4	58	1	1	3	0.583	4.497
11	2	1	2	1	1	47	5	1	13	59	0	3	1	3.591	0.419
12	2	1	3	2	2	46	6	8	6	59	0	4	0	6.909	2.342
13	2	2	1	2	3	46	6	10	4	59	0	0	4	−10.939	8.237
14	2	2	2	3	1	49	3	11	3	59	0	4	0	8.354	2.342
15	2	2	3	1	2	46	6	10	4	59	0	0	4	−10.939	8.237
16	2	3	1	3	2	50	2	2	12	59	0	0	4	4.120	8.237
17	2	3	2	1	3	50	2	4	10	57	2	0	4	1.784	4.841
18	2	3	3	2	1	44	8	6	8	59	0	3	1	−5.697	0.419

n_{00} = the number of correct design parts detected accurately as correct design parts,

n_{01} = the number of correct design parts detected by mistake as incorrect design parts,

n_{10} = the number of incorrect design parts detected by mistake as correct design parts,

n_{11} = the number of incorrect design parts detected accurately as incorrect design parts,

where two kinds of error rate are defined by

$$p = \frac{n_{01}}{n_0}, \tag{2.2}$$

$$q = \frac{n_{10}}{n_1}, \tag{2.3}$$

Considering the two kinds of error rate, p and q, we can derive the *standard error rate*, p_0, [4] as

$$p_0 = \frac{1}{1 + \sqrt{\left(\frac{1}{p} - 1\right)\left(\frac{1}{q} - 1\right)}}. \tag{2.4}$$

Then, the signal-to-noise ratio based on (2.4) is defined by (see [4])

$$\eta_0 = -10\log_{10}\left\{\frac{1}{(1-2p_0)^2} - 1\right\}. \tag{2.5}$$

The standard error rate, p_0, can be obtained from transforming (2.5) by using the signal-to-noise ratio of each control factor as

$$p_0 = \frac{1}{2}\left\{1 - \frac{1}{\sqrt{10^{(-\frac{\eta_0}{10})}+1}}\right\}. \tag{2.6}$$

The method of experimental design based on orthogonal-arrays is a special one that require only a small number of experimental trials to help us discover main factor effects. On traditional researches [7, 11], the design of experiment has been conducted by using orthogonal-array $L_{12}(2^{11})$. However, since the orthogonal-array $L_{12}(2^{11})$ has only two levels for grasp of factorial effect to the human factors experiment, the middle effect between two levels can not be measured. Thus, in order to measure it, we adopt the orthogonal-array $L_{18}(2^1 \times 3^7)$ can lay out one factor with 2 levels (1, 2) and 7 factors with 3 levels (1, 2, 3) as shown in Table 2.3, and dispense with $2^1 \times 3^7$ trials by executing experimental independent 18 experimental trials each other. For example, as for the experimental trial No. 10, we executed the design-review work under the conditions A_2, B_1, C_1, D_3, and E_3, and obtained the computed SNR's as 0.583 (dB) for the descriptive-design faults from the observed values $n_{00} = 52$, $n_{01} = 0$, $n_{10} = 10$, and $n_{11} = 4$, and as 4.497 (dB) for the symbolical-design faults from the observed values $n_{00} = 58$, $n_{01} = 1$, $n_{10} = 1$, and $n_{11} = 3$.

2.2.4 Investigation of Analysis Results

We analyze simultaneous effects of outside factor R and inside human factors A, B, C, D, and E. As a result of the analysis of variance by taking account of correlation among inside and outside factors discussed in Sect. 2.2.2, we can obtain Table 2.4. There are two kinds of errors in the analysis of variance: e_1 is the error among experiments of the inside factors, and e_2 the mutual correlation error between e_1 and the outside factor. In this analysis, since there was no significant effect by performing F-test for e_1 with e_2, F-test for all factors was performed by e_2. As the result, the significant human factors such as the degree of understanding of the design-method (Factor C), the degree of understanding of requirement-specification (Factor D), and the classification of detected faults (Factor R) were recognized. Figure 2.3 shows the factor effect for each level in the significant factors which affect design-review work.

As a result of analysis, in the inside factors, only Factor C and D are significant and the inside and outside factors are not mutually interacted. That is, it turns out that the

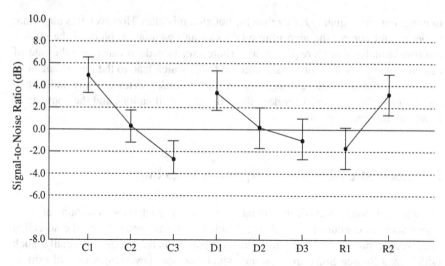

Fig. 2.3 The estimation of significant factors with correlation among inside and outside factors

Table 2.4 The result of analysis of variance by taking account of correlation among inside and outside factors

Factor	f	S	V	F_0	$\rho(\%)$
A	1	37.530	37.530	2.497	3.157
B	2	47.500	23.750	1.580	3.995
C	2	313.631	156.816	10.435**	26.380
D	2	137.727	68.864	4.582*	11.584
E	2	4.684	2.342	0.156	0.394
A × B	2	44.311	22.155	1.474	3.727
e_1	6	38.094	6.460	0.422	3.204
R	1	245.941	245.941	16.366**	20.686
A × R	1	28.145	28.145	1.873	2.367
B × R	2	78.447	39.224	2.610	6.598
C × R	2	36.710	18.355	1.221	3.088
D × R	2	9.525	4.763	0.317	0.801
E × R	2	46.441	23.221	1.545	3.906
e_2	8	120.222	15.028	3.870	10.112
T	35	1188.909			100.0

*:5 % level of significant
**:1 % level of significant

reviewers with the high degree of understanding of the design-method and the high degree of understanding of requirement-specification can review the design-specification efficiently regardless of the classification of detected faults. Moreover, the result that outside factor R is highly significant, and the descriptive-design faults are detected less than the symbolic-design faults, can be obtained. That is, although it is a natural result, it is difficult to detect and correct the algorithmic faults which lead to improvement in quality rather than the notation mistakes. However, it is important to detect and correct the algorithmic faults as an essential problem of the quality improvement for design-review work. Therefore, in order to increase the rate of detection and correction of the algorithmic faults which lead to the improvement of quality, it is required before design-review work to make reviewers fully understand the design techniques used for describing design-specifications and the contents of requirement-specifications.

2.3 Stochastic Differential Equation Modeling

Software development environment has been changing into new development paradigms such as concurrent distributed development environment and the so-called open source project by using network computing technologies. Especially, such OSS (Open Source Software) systems which serve as key components of critical infrastructures in the society are still ever-expanding now [12].

The successful experience of adopting the distributed development model in such open source projects includes GNU/Linux operating system, Apache Web server, and so on [12]. However, the poor handling of the quality and customer support prohibit the progress of OSS. We focus on the problems in the software quality, which prohibit the progress of OSS.

Especially, SRGM's [6, 13] have been applied to assess the reliability for quality management and testing-progress control of software development. On the other hand, the effective method of dynamic testing management for new distributed development paradigm as typified by the open source project has only a few presented [14–17]. In case of considering the effect of the debugging process on entire system in the development of a method of reliability assessment for OSS, it is necessary to grasp the situation of registration for bug tracking system, degree of maturation of OSS, and so on.

In this chapter, we focus on an open source solution developed under several OSS's. We discuss a useful software reliability assessment method in open source solution as a typical case of next-generation distributed development paradigm. Especially, we propose a software reliability growth model based on stochastic differential equations in order to consider the active state of the open source project and the component collision of OSS. Then, we assume that the software failure intensity depends on the time, and the software fault-reporting phenomena on the bug tracking system keep an irregular state. Also, we analyze actual software fault-count data to show numerical examples of software reliability assessment for the open source solution. Moreover, we compare our model with the conventional model based on stochastic differential equations in terms of goodness-of-fit for actual data. Then, we show that the proposed model can assist improvement of quality for an open source solution developed under several OSS's.

2.3.1 Stochastic Differential Equation Model

Let $S(t)$ be the number of detected faults in the open source solution by testing time $t(t \geq 0)$. Suppose that $S(t)$ takes on continuous real values. Since latent faults in the open source solution are detected and eliminated during the operational phase, $S(t)$ gradually increases as the operational procedures go on. Thus, under common assumptions for software reliability growth modeling, we consider the following linear differential equation:

$$\frac{dS(t)}{dt} = \lambda(t)S(t), \tag{2.7}$$

where $\lambda(t)$ is the intensity of inherent software failures at operational time t and is a non-negative function.

Generally, it is difficult for users to use all functions in open source solution, because the connection state among open source components is unstable in the testing-phase of open source solution. Considering the characteristic of open source solution, the software fault-report phenomena keep an irregular state in the early stage of testing phase. Moreover, the addition and deletion of software components are repeated under the development of an OSS system, i.e. we consider that the software failure intensity depends on the time.

Therefore, we suppose that $\lambda(t)$ and $\mu(t)$ have the irregular fluctuation. That is, we extend (2.7) to the following stochastic differential equation (SDE) [18, 19]:

$$\frac{dS(t)}{dt} = \{\lambda(t) + \sigma\mu(t)\gamma(t)\}S(t), \tag{2.8}$$

where σ is a positive constant representing a magnitude of the irregular fluctuation, $\gamma(t)$ a standardized Gaussian white noise, and $\mu(t)$ the collision level function of open source component.

We extend (2.8) to the following stochastic differential equation of an Ito type:

$$dS(t) = \left\{\lambda(t) + \frac{1}{2}\sigma^2\mu(t)^2\right\}S(t)dt + \sigma\mu(t)S(t)d\omega(t), \tag{2.9}$$

where $\omega(t)$ is a one-dimensional Wiener process which is formally defined as an integration of the white noise $\gamma(t)$ with respect to time t. The Wiener process is a Gaussian process and it has the following properties:

$$\Pr[\omega(0) = 0] = 1, \tag{2.10}$$

$$E[\omega(t)] = 1, \tag{2.11}$$

$$E[\omega(t)\omega(t')] = \mathrm{Min}[t, t']. \tag{2.12}$$

By using Ito's formula [18, 19], we can obtain the solution of (2.8) under the initial condition $S(0) = \nu$ as follows [20]:

$$S(t) = v \cdot \exp\left(\int_0^t \lambda(s)ds + \sigma\mu(t)\omega(t)\right),\qquad(2.13)$$

where v is the number of detected faults for the previous software version. Using solution process $S(t)$ in (2.13), we can derive several software reliability measures.

Moreover, we define the intensity of inherent software failures, $\lambda(t)$, and the collision level function, $\mu(t)$, as follows:

$$\int_0^t \lambda(s)ds = (1 - \exp[-\alpha t]),\qquad(2.14)$$

$$\mu(t) = \exp[-\beta t],\qquad(2.15)$$

where α is an acceleration parameter of the intensity of inherent software failures, and β the growth parameter of the open source project.

2.3.2 Method of Maximum-Likelihood

In this section, the estimation method of unknown parameters α, β and σ in (2.13) is presented. Let us denote the joint probability distribution function of the process $S(t)$ as

$$P(t_1, y_1; t_2, y_2; \cdots; t_K, y_K) \equiv \Pr[S(t_1) \le y_1, \cdots, S(t_K) \le y_K \mid S(t_0) = v],\qquad(2.16)$$

where $S(t)$ is the cumulative number of faults detected up to the operational time $t (t \ge 0)$, and denote its density as

$$p(t_1, y_1; t_2, y_2; \cdots; t_K, y_K) \equiv \frac{\partial^K P(t_1, y_1; t_2, y_2; \cdots; t_K, y_K)}{\partial y_1 \partial y_2 \cdots \partial y_K}.\qquad(2.17)$$

Since $S(t)$ takes on continuous values, we construct the likelihood function l for the observed data $(t_k, y_k)(k = 1, 2, \cdots, K)$ as follows:

$$l = p(t_1, y_1; t_2, y_2; \cdots; t_K, y_K).\qquad(2.18)$$

For convenience in mathematical manipulations, we use the following logarithmic likelihood function:

$$L = \log l.\qquad(2.19)$$

The maximum-likelihood estimates α^*, β^*, and σ^* are the values making L in (2.19) maximize. These can be obtained as the solutions of the following simultaneous likelihood equations [20]:

$$\frac{\partial L}{\partial \alpha} = \frac{\partial L}{\partial \beta} = \frac{\partial L}{\partial \sigma} = 0.\qquad(2.20)$$

2.3.3 Expected Number of Detected Faults

We consider the expected number of faults detected up to operational time t. The density function of $\omega(t)$ is given by:

$$f(\omega(t)) = \frac{1}{\sqrt{2\pi t}} \exp\left\{-\frac{\omega(t)^2}{2t}\right\}. \tag{2.21}$$

Information on the cumulative number of detected faults in the OSS system is important to estimate the situation of the progress on the software operational procedures. Since it is a random variable in our model, its expected value and variance can be useful measures. We can calculate the expected number of faults detected up to time t from (2.13) as follows [20]:

$$E[S(t)] = \nu \cdot \exp\left(\int_0^t \lambda(s)ds + \frac{\sigma^2\mu(t)^2}{2}t\right). \tag{2.22}$$

2.3.4 Numerical Illustrations

We focus on a large scale open source solution based on the Apache HTTP Server [21], Apache Tomcat [22], MySQL [23] and JSP (JavaServer Pages). The fault-count data used in this chapter are collected in the bug tracking system on the website of each open source project. The estimated expected cumulative number of detected faults in (2.22) is shown in Fig. 2.4. Also, the sample path of the estimated numbers of detected faults in (2.13) is shown in Fig. 2.5, approximately.

We show the reliability assessment results for the other SDE model in terms of the performance evaluation of our model. The sample path of the estimated cumulative numbers of detected faults in the conventional SDE model for OSS [24] are shown in Fig. 2.6. Also, Fig. 2.7 is the sample path of the estimated numbers of remaining faults in the conventional SDE model [25]. From Figs. 2.6 and 2.7, we have found that the magnitude of the irregular fluctuation in the early phase of the proposed model is larger than those of the conventional SDE models, i.e. the irregular fluctuation in the proposed model depends on the time. Then, for the large scale open source solution [26, 27], we may utilize the proposed model for assisting improvement of quality, in which it can describe actual fault-detection phenomena.

2.4 Discrete NHPP Modeling

In recent researches, Satoh [28] proposed a discrete Gompertz curve model, and Satoh and Yamada [29] suggested parameter estimation procedures for software reliability assessment of a discrete logistic curve model, and compared the these models by

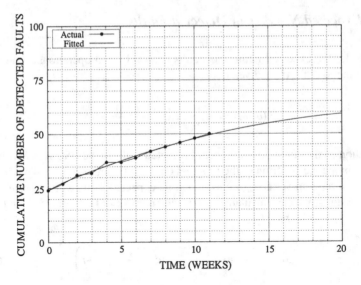

Fig. 2.4 The estimated cumulative number of detected faults, $E[S(t)]$

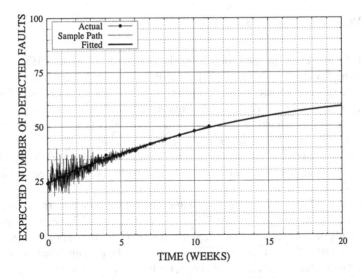

Fig. 2.5 The estimated path of the estimated number of detected faults

using a new proposed criterion. They reported that the discrete models as statistical data analysis models enable us to obtain accurate parameter estimates even with a small amount of observed data for particular applications.

In this section, we discuss the discrete nonhomogeneous Poisson process (NHPP) models [6] derived by employing a difference method which conserves the gauge invariance from above results and high applicability of NHPP models point of view.

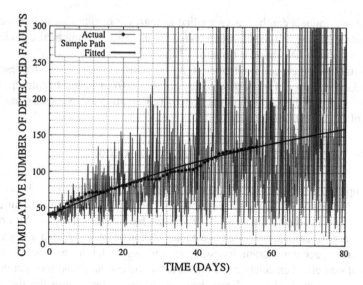

Fig. 2.6 The sample path of the estimated cumulative number of detected faults for the conventional SDE model for OSS

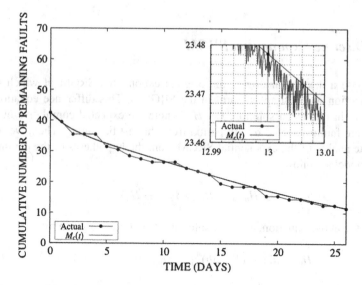

Fig. 2.7 The sample path of the estimated number of remaining faults for the conventional SDE model

The discrete NHPP models, that is, the discrete exponential SRGM and the discrete inflection S-shaped SRGM, have exact solutions. The difference equations and their exact solutions tend to the differential equations and their exact solutions. Therefore, the proposed models conserve the characteristics of the continuous NHPP models.

The proposed models can be easily applied to regression equations to get accurate parameter estimates, and have more advantages in terms of numerical calculations than the maximum-likelihood estimation [30].

We assume a discrete counting process $\{N_n, n \geq 0\}(n = 0, 1, 2, \cdots)$ representing the cumulative number of faults detected by nth period from the test beginning. Then, the NHPP model with mean value function D_n representing the expected cumulative number of faults is formulated by

$$\Pr\{N_n = x\} = \frac{[D_n]^x}{x!}\exp[-D_n] \quad (n, x = 0, 1, 2, \cdots). \tag{2.23}$$

We employ a difference method which conserves the gauge invariance because the proposed discrete NHPP models have to conserve the characteristic of the continuous NHPP models, i.e. the continuous NHPP models have exact solutions. With regard to parameter estimations, the difference equations can be easily applied to regression equations to get accurate parameter estimates, and this models have some advantages in terms of numerical calculations. Therefore, we can estimate unknown parameters by adopting the method of ordinary least-square procedures from the regression equations.

2.4.1 Discrete Exponential SRGM

We propose a discrete analog of the original exponential SRGM of which mean value function is the simplest form in the SRGM's. This difference equation for this model has an exact solution. Let H_n denote the expected cumulative number of software faults detected by nth period from the test beginning. Then, we derive a discrete analog of the exponential SRGM from the hypotheses of the continuous NHPP model as follows:

$$H_{n+1} - H_n = \delta b(a - H_n). \tag{2.24}$$

Solving the above equation, an exact solution H_n in (2.24) is given by

$$H_n = a[1 - (1 - \delta b)^n] \quad (a > 0, \; 0 < b < 1), \tag{2.25}$$

where δ represents the constant time-interval, a the expected total number of potential software failures occurred in an infinitely long duration or the expected initial fault content, and b the fault-detection rate per fault. As $\delta \rightarrow 0$, (2.25) converges to an exact solution of the original exponential SRGM which is described by the differential equation.

We can derive a regression equation from (2.24) to estimate the model parameters. The regression equation is obtained as

$$Y_n = A + BH_n, \tag{2.26}$$

where

$$\begin{cases} Y_n = H_{n+1} - H_n \\ A = \delta ab \\ B = -\delta b. \end{cases} \tag{2.27}$$

Using (2.26), we can estimate $\hat{A}$ and $\hat{B}$ by using the observed data, which are the estimates of A and B. Therefore, we can obtain the parameter estimates $\hat{a}$ and $\hat{b}$ from (2.27) as follows:

$$\begin{cases} \hat{a} = -\hat{A}/\hat{B} \\ \hat{b} = -\hat{B}/\delta. \end{cases} \tag{2.28}$$

Y_n in (2.26) is independent of δ because δ is not used in calculating Y_n in (2.26). Hence, we can obtain the same parameter estimates of $\hat{a}$ and $\hat{b}$, respectively, when we choose any value of δ.

2.4.2 Discrete Inflection S-Shaped SRGM

We also propose a discrete analog of the original inflection S-shaped SRGM which is the continuous one. Let I_n denote the expected cumulative number of software faults detected by nth period from the test beginning. Then, we can derive a discrete analog of the inflection S-shaped SRGM from the hypotheses of the continuous NHPP model as follows:

$$I_{n+1} - I_n = \delta abl + \frac{\delta b(1 - 2l)}{2}[I_n + I_{n+1}] - \frac{\delta b(1 - l)}{a} I_n I_{n+1}. \tag{2.29}$$

Solving the above difference equation, an exact solution I_n in (2.29) is given by

$$I_n = \frac{a\left[1 - \left(\frac{1 - \frac{1}{2}\delta b}{1 + \frac{1}{2}\delta b}\right)^n\right]}{1 + c\left(\frac{1 - \frac{1}{2}\delta b}{1 + \frac{1}{2}\delta b}\right)^n} \quad (a > 0, \ 0 < b < 1, \ c > 0, \ 0 \leq l \leq 1), \tag{2.30}$$

where δ represents the constant time-interval, a the expected total number of potential software failures occurred in an infinitely long duration or the expected initial fault content, b the fault-detection rate per fault, and c the inflection parameter. The inflection parameter is specified as follows: $c = (1 - l)/l$ where l is the inflection rate which indicates the ratio of the number of detectable faults to the total number of faults in the software system. As $\delta \to 0$, (2.30) converges to an exact solution of the original inflection S-shaped SRGM which is described by the differential equation.

Defining the difference operator as

$$\Delta I_n \equiv \frac{I_{n+1} - I_n}{\delta}. \tag{2.31}$$

We show that the inflection point occurs when

$$\bar{n} = \begin{cases} <n^*> & (if \, \Delta I_{<n^*>} \geq \Delta I_{<n^*>+1}) \\ <n^*> +1 & (otherwise), \end{cases} \tag{2.32}$$

where

$$n^* = -\frac{\log c}{\log\left(\frac{1-\frac{1}{2}\delta b}{1+\frac{1}{2}\delta b}\right)} - 1, \tag{2.33}$$

$$<n^*> = \{n|\max(n \leq n^*), \, n \in Z\}. \tag{2.34}$$

Moreover, we define t^* as

$$t^* = n^* \delta. \tag{2.35}$$

When n^* is an integer, we can show that t^* converges the inflection point of the inflection S-shaped SRGM which is described by the differential equation as $\delta \to 0$ as follows:

$$t^* = -\delta \frac{\log c}{\log\left(\frac{1-\frac{1}{2}\delta b}{1+\frac{1}{2}\delta b}\right)} - \delta \to \frac{\log c}{b} \text{ as } \delta \to 0. \tag{2.36}$$

By the way, the inflection S-shaped SRGM is regarded as a Riccati equation. Hirota [31] proposed a discrete Riccati equation which has an exact solution. A Bass model [32] which forecasts the innovation diffusion of products is also a Riccati equation. Satoh [33] proposed a discrete Bass model which can overcome the shortcomings of the ordinary least-square procedures in the continuous Bass model.

We can derive a regression equation to estimate the model parameters from (2.29). The regression equation is obtained as

$$Y_n = A + BK_n + CL_n, \tag{2.37}$$

where

$$\begin{cases} Y_n = I_{n+1} - I_n \\ K_n = I_n + I_{n+1} \\ L_n = I_n I_{n+1} \\ A = \delta abl \\ B = \delta b(1 - 2l)/2 \\ C = -\delta b(1 - l)/a. \end{cases} \tag{2.38}$$

Using (2.37), we can estimate $\hat{A}$, $\hat{B}$, and $\hat{C}$ by using the observed data, which are the estimates of A, B, and C, respectively. Therefore, we can obtain the parameter estimates $\hat{a}$, $\hat{b}$, and $\hat{l}$ from (2.38) as follows:

$$\begin{cases} \hat{a} = \hat{A}/(\sqrt{\hat{B}^2 - \hat{A}\hat{C}} - \hat{B}) \\ \hat{b} = 2\sqrt{\hat{B}^2 - \hat{A}\hat{C}}/\delta \\ \hat{l} = (1 - \hat{B}/\sqrt{\hat{B}^2 - \hat{A}\hat{C}})/2. \end{cases} \tag{2.39}$$

Y_n, K_n, and L_n in (2.37) are independent of δ because δ is not used in calculating Y_n, K_n, and L_n in (2.37). Hence, we can obtain the same parameter estimates $\hat{a}$, $\hat{b}$, and $\hat{l}$, respectively, when we choose any value of δ.

2.4.3 Model Comparisons

We show the result of goodness-of-fit comparisons in this section. We compare the four discrete models by using four data sets (DS1–DS4) observed in actual software testing. The four discrete models are as follows: two discrete NHPP models that were discussed in Sects. 2.4.1 and 2.4.2, a discrete logistic curve model [29, 30], and a discrete Gompertz curve model [28]. The data sets of DS1 and DS2 indicate exponential growth curves, and those of DS3 and DS4 indicate S-shaped growth curves, respectively. We employ the predicted relative error [6], the mean square errors (MSE) [6], and Akaike's Information Criterion (AIC) [6] as criteria of the model comparison in this section.

The predicted relative error is a useful criterion for indicating the relative errors between the predicted number of faults discovering by termination time of testing by using the part of observed data from the test beginning and the observed number of faults discovering by the termination time. Let $R_e[t_e]$ denote the predicted relative error at arbitrary testing time t_e. Then, the predicted relative error is given by

$$R_e[t_e] = \frac{\hat{y}(t_e; t_q) - q}{q}, \tag{2.40}$$

where $\hat{y}(t_e; t_q)$ is the estimated value of the mean value function at the termination time t_q by using the observed data by the arbitrary testing time $t_e (0 \leq t_e \leq t_q)$, and q is the observed cumulative number of faults detected by the termination time. We show Figs. 2.8, 2.9, 2.10 which are the results of the model comparisons based on the predicted relative error for DS1 and DS3. MSE is obtained by using the sum of squared errors between the observed and estimated cumulative numbers of detected faults, y_k and $\hat{y}(t_k)$ during $(0, t_k]$, respectively. Getting N data pairs $(t_k, y_k)(k = 1, 2, \cdots, N)$, MSE is given by

$$\text{MSE} = \frac{1}{N} \sum_{k=1}^{N} [y_k - \hat{y}(t_k)]^2, \tag{2.41}$$

where $\hat{y}(t_k)$ denote the estimated value of the expected cumulative number of faults by using exact solutions of each model by the arbitrary testing time $t_k (k = 1, 2, \cdots, N)$. Table 2.5 shows the result of model comparison based on MSE for the each model. From Table 2.5, we conclude that the discrete inflection S-shaped SRGM fits better to all data sets except for DS2. However, the result of model comparison based on MSE depends on the number of model parameters of each models, e.g., the discrete exponential SRGM has two parameters and the discrete inflection S-shaped one has three parameters. Therefore, as a criterion of goodness-of-fit comparison for the two discrete models, i.e. the discrete exponential SRGM and the discrete inflection S-shaped one, we adopt the value of AIC. Table 2.6 shows the result of model comparison based on AIC. From Table 2.6, we can validate the above evaluation for MSE.

From these three results of goodness-of-fit comparison, we conclude that the discrete exponential SRGM is more useful model for software reliability assessment for the observed data which indicate an exponential growth curve, and the discrete

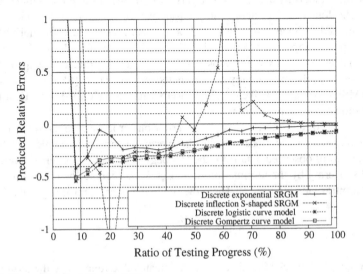

Fig. 2.8 The predicted relative for DS1

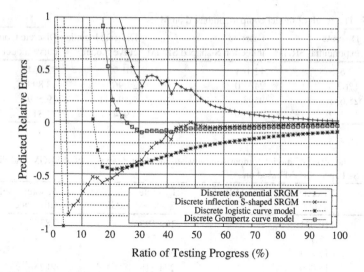

Fig. 2.9 The predicted relative error for DS3

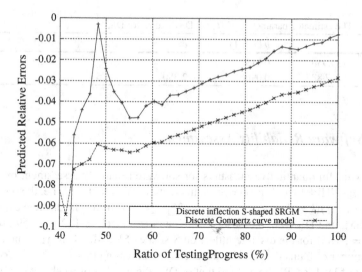

Fig. 2.10 The model comparison based on the predicted relative error for DS3 focusing on the discrete Gompertz curve model and the discrete inflection S-shaped SRGM

inflection S-shaped SRGM is more useful one for assessment after 60 % of the testing progress ratio for the observed data which indicate an S-shaped growth curve.

Table 2.5 The result of model comparison based on MSE

Data set	Discrete exponential SRGM	Discrete inflection S-shaped SRGM	Discrete logistic curve model	Discrete Gompertz curve model
DS1	39.643	12.141	101.92	72.854
DS2	1762.5	2484.0	27,961	13,899
DS3	25,631	9598.1	1,49,441	19,579
DS4	11,722	438.59	4,9741	27,312

Table 2.6 The result of model comparison between the discrete exponential SRGM and discrete inflection S-shaped SRGM based on AIC

Data set	Discrete expo-nential SRGM	Discrete inflection S-shaped SRGM	Absolute value of difference
DS1	110.031	109.195	0.836
DS2	115.735	118.752	3.017
DS3	617.434	606.132	11.30
DS4	315.069	274.818	40.25

Table 2.7 The estimated parameters of $\hat{H}_n$ for DS1 and $\hat{I}_n$ for DS3

	$\hat{a}$	$\hat{b}\,(\delta = 1)$	$\hat{c}$	n^*	$< n^* >$	$\bar{n}$
Hn	139.956	0.113				
In	5217.88	0.0906	2.350	8.385	8	9

2.4.4 Software Reliability Assessment

We show useful quantitative measures for software reliability assessment by using the discrete NHPP models proposed in this section. We adopt DS1, i.e. the observed 25 pairs $(t_k, y_k)(k = 1, 2, \cdots, 25\,;\ t_{25} = 25,\ y_{25} = 136)$ for the discrete exponential SRGM, and DS3, i.e. the observed 59 pairs $(t_k, y_k)(k = 1, 2, \cdots, 59\,;\ t_{59} = 59,\ y_{59} = 5186)$ for the discrete inflection S-shaped SRGM, where y_k is the cumulative number of faults detected by the execution of testing time t_k. The observation time unit of DS1 is CPU hours, and that of DS3 the number of weeks. We show the estimated mean value functions of H_n in (2.25) and I_n in (2.30) in Figs. 2.11 and 2.12, respectively, where several quantities are shown in Table 2.7.

We can derive the software reliability function which is a useful measure for software reliability assessment. The software reliability function is obtained by (2.23) as follows:

$$R(n, h) \equiv \Pr\{N_{n+h} - N_n = 0|N_n = x\}$$
$$= \exp[-\{D_{n+h} - D_n\}]. \tag{2.42}$$

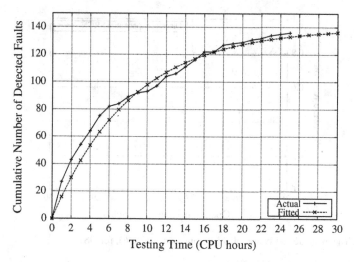

Fig. 2.11 The estimated discrete mean value function, $\hat{H}_n$, for DS1.

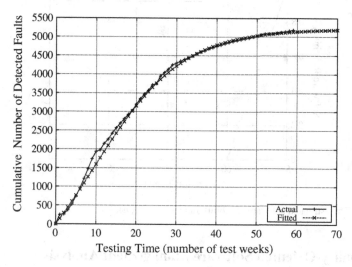

Fig. 2.12 The estimated discrete mean value function, $\hat{I}_n$, for DS3

Letting $\delta = 1$, the software reliability function for H_n after the termination time $n = 25$ (CPU hours), and for I_n after the termination time $n = 59$ (weeks), are shown in Figs. 2.13 and 2.14, respectively. After releasing the software systems at these time points, assuming that the software users operate these software systems under the same environment as the software testing one, we can estimate the software reliability $\hat{R}(25, 1.0)$ for H_n to be about 0.46. Also, we can estimate one $\hat{R}(59, 0.1)$ for I_n to be about 0.48.

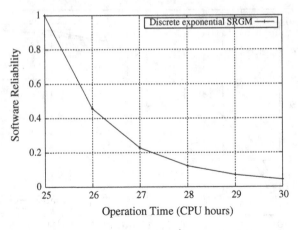

Fig. 2.13 The estimated software reliability function, $\hat{R}(25, h)$, for DS1

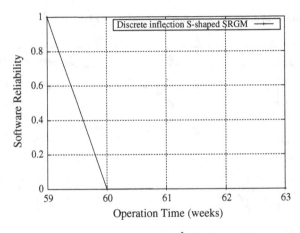

Fig. 2.14 The estimated software reliability function, $\hat{R}(59, h)$, for DS3

2.5 Quality-Oriented Software Management Analysis

In this section, firstly, we conduct multivariate linear analyses by using process monitoring [34] data, derive effective process factors affecting the final products' quality, and discuss the significant process factors with respect to software management measures of QCD [35, 36]. Then, we analyze actual process monitoring data, based on the derivation procedures of a process improvement model, i.e. software management model [37, 38] (as shown in Fig. 2.15). Then, we discuss project management on the significant process factors affecting the QCD measures, and show their effect on them. Secondly, we analyze the process monitoring data in a viewpoint of software reliability measurement and assessment in the process monitoring activities.

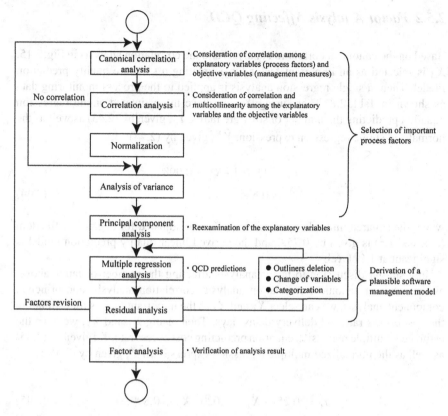

Fig. 2.15 Derivation procedures of software management model

2.5.1 Process Monitoring Data

We predict software management measures of QCD by using the process monitoring data as shown in Table 2.8. Five variables measured in terms of the number of faults (QCD problems) detected through the process monitoring, i.e. the contract review, the development planning review, the design completion review, the test planning review, and the test completion review phases are used as explanatory variables. The observed values of these five factors are normalized by each project development size (KLOC) in this section. Three variables, i.e. the number of faults detected during customer acceptance testing, the cost excess rate, and the number of delivery-delay days, are used as objective variables.

2.5.2 Factor Analysis Affecting QCD

Based on the canonical correlation analysis and the correlation analysis in Fig. 2.15, X_3 is selected as an important factor for estimating a software quality prediction model. Then, a single regression analysis is applied to the process monitoring data as shown in Table 2.8. Then, using X_3, we have the estimated single regression equation predicting the number of software faults, $\hat{Y}_q$, given by (2.43) as well as the normalized single regression expression, $\hat{Y}_q^N$, given by (2.44):

$$\hat{Y}_q = 11.761 \cdot X_3 + 0.998, \tag{2.43}$$

$$\hat{Y}_q^N = 0.894 \cdot X_3, \tag{2.44}$$

where the squared multiple correlation coefficient adjusted for degrees of freedom (adjusted R^2) is given by 0.758, and the derived linear quality prediction model is significant at 1 % level.

In a similar discussion to factor analysis affecting the number of faults above, as the result of canonical correlation analysis, correlation analysis, and principal component analysis, we can select X_1 and X_5 as the important factors for estimating the cost excess rate and delivery-delay days. Then, using X_1 and X_5, we have the estimated multiple regression equation predicting cost excess rate, $\hat{Y}_c$, given by (2.45) as well as the normalized multiple regression expression, $\hat{Y}_c^N$, given by (2.46):

$$\hat{Y}_c = 0.253 \cdot X_1 + 1.020 \cdot X_5 + 0.890, \tag{2.45}$$

$$\hat{Y}_c^N = 0.370 \cdot X_1 + 0.835 \cdot X_5, \tag{2.46}$$

where the adjusted R^2 is given by 0.917, and the derived cost excess prediction model is significant at 1 % level.

By the same way of the cost excess rate, using X_1 and X_5, we have the estimated multiple regression equation predicting the number of delivery-delay days, $\hat{Y}_d$, given by (2.47) as well as the normalized multiple regression expression, $\hat{Y}_d^N$, given by (2.48):

$$\hat{Y}_d = 24.669 \cdot X_1 + 55.786 \cdot X_5 - 9.254, \tag{2.47}$$

$$\hat{Y}_d^N = 0.540 \cdot X_1 + 0.683 \cdot X_5, \tag{2.48}$$

where the adjusted R^2 is given by 0.834, and the derived delivery-delay prediction model is significant at 5 % level.

Table 2.8 Process monitoring data

Project no	Contract review (X_1) Number of faults per development size	Development planning (X_2) Number of faults per development size	Design completion review (X_3) Number of faults per development size	Test planning review (X_4) Number of faults per development size	Test completion review (X_5) Number of faults per development size	Quality (Y_q) Number of faults detected during acceptance testing	Cost (Y_c) Cost excess rate	Delivery (Y_d) Number of delivery-delay days
1	0.591	1.181	0.295	0.394	0.394	4	1.456	28
2	0.323	0.645	0	0.108	0.108	1	1.018	3
3	0.690	0.345	0	0.345	0	0	1.018	4
4	0.170	0.170	0	0.085	0	2	0.953	0
5	0.150	0.451	0.301	0.075	0.075	5	1.003	0
6	1.186	0.149	0	0.037	0.037	0	1	−8
7	0.709	0	0	0	0	2	1.119	12

2.5.3 Analysis Results of Software Management Models

We have derived software management models by applying the methods of multivariate linear analysis to actual process monitoring data. Quantitative evaluation based on the derived prediction models about final product quality, cost excess, and delivery-delay, has been conducted with high accuracy. Then, it is so effective to promote software process improvement under PDCA (Plan, Do, Check, Act) management cycle by using the derivation procedures of software management models as shown in Fig. 2.15.

Further, the design completion review has an important impact on software quality. Then, it is possible to predict software product quality in the early-stage of software development project by using the result of the design completion review in process monitoring activities.

Next, the contract review and the test completion review processes have important impact on the cost excess rate and the delivery-delay days. That is, it is difficult to predict cost excess and delivery-delay measures at the early-stage of software development project, and it is found that the cost excess and delivery-delay measures can be predicted according to the same process monitoring factors.

2.5.4 Implementation of Project Management

2.5.4.1 Continuous Process Improvement

From the result of software management model analyses and factor analyses, it is found that the contract review has an important relationship with cost and delivery measures. Then, in order to improve the cost excess and delivery-delay, we perform suitable project management for the problems detected in the contract review.

The project management practices to be performed for the important problems detected in the contract review are:

- Early decision of the specification domain.
- Improvement of requirement specification technologies.
- Early decision of development schedule.
- Improvement of project progress management.
- Improvement of testing technology.

As a result of carrying out project management and continuous process improvement, the relationship between the risk ratio measured at the initial stage of a project and the amount of problem solving effort (man-day) in the contract review become as shown in Fig. 2.16 where Projects 8–21 were monitored under process improvement based on the analysis results for Projects 1–7, and the risk ratio is given by

$$R = \sum_i \{risk\ item(i) \times weight(i)\}. \tag{2.49}$$

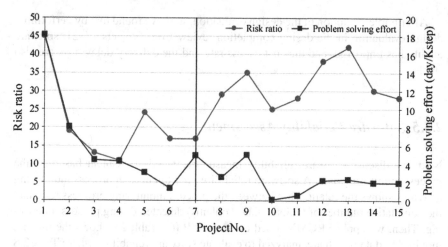

Fig. 2.16 Relationship between risk ratio and problem solving effort

In (2.49), the risk estimation checklist has weight(i) in each risk item(i), and the risk ratio ranges between 0 and 100 points. Project risks are identified by interviewing based on the risk estimation checklist. From the identified risks, the risk ratio of a project is calculated by (2.49).

From Fig. 2.16, it is found that by performing suitable project management for the important problems in the contract review from Projects 8–15, the problem can be solved at the early stage of software project even if the risk ratio is high.

2.5.4.2 Implementation of Design Quality Evaluation

In a similar fashion to cost and delivery measures, it is found that the design completion review has an important relationship with software quality. Then, in order to improve software quality, we decide to perform suitable project management called design evaluation in the design completion review.

The design evaluation assesses the following items based on the risk estimation checklist by the project manager, the designer, and the members of quality control department. Through the following design evaluation, we have to judge if the development can proceed to the next stage:

- After the requirements analysis, how many requirements are included in the requirement specifications? Are the requirements (function requirements and non-function requirements) suitably-defined?
- After the elementary design, has the requirements (function requirements and non-function requirements) been taken over from the user requirements to the design documents without omission about the description items in the requirement specification?
- As for elementary design documents, is the elementary design included?

After implementation of the design evaluation, we have found that by performing design evaluation in the design completion review from Projects 17–21, software quality has improved, and the cost excess rate and the delivery-delay days are also stable.

2.5.5 Software Reliability Assessment

Next, we discuss software reliability measurement and assessment based on the process monitoring data. A software reliability growth curve in process monitoring activities shows the relationship between the process monitoring progress ratio and the cumulative number of faults (QCD problems) detected during process monitoring. Then, we apply SRGM's based on an NHPP [6]. Table 2.9 shows the process monitoring data which are analyzed to evaluate software reliability, where Table 2.8 is derived from Table 2.9 for Projects 1–7, and Projects 8–21 were monitored under process improvement based on the analysis results for Projects 1–7. However, the collected process monitoring data have some missing values in metrics. Therefore, we apply collaborative filtering to the observed data to complement the missing values for assessing software reliability. The under-lined values in Table 2.9 are the metrics values complemented by collaborative filtering.

We discuss software reliability growth modeling based on an NHPP because an analytic treatment of it is relatively easy. Then, we choose the process monitoring progress ratio as the alternative unit of testing-time by assuming that the observed data for testing-time are continuous.

In order to describe a fault-detection phenomenon at processing monitoring progress ratio t $(t \geq 0)$, let $\{N(t), t \geq 0\}$ denote a counting process representing the cumulative number of faults detected up to progress ratio t. Then, the fault-detection phenomenon can be described as follow:

$$\Pr\{N(t) = n\} = \frac{\{H(t)\}^n}{n!} \exp[-H(t)] \ (n = 0, 1, 2, \cdots), \quad (2.50)$$

where $H(t)$ represents the expected value of $N(t)$ called a mean value function of the NHPP. $\Pr\{A\}$ in (2.50) means the probability of event A. In this section, we apply three NHPP models [6], i.e. the exponential SRGM, the delayed S-shaped SRGM, and the logarithmic Poisson execution time model.

Software reliability assessment measures play an important role in quantitative software reliability assessment based on an SRGM. The expected number of remaining faults, $n(t)$, represents the number of faults latent in the software system by arbitrary testing-time t, and is formulated as

$$n(t) \equiv \mathrm{E}[N(\infty) - N(t)] = \mathrm{E}[N(\infty)] - H(t), \quad (2.51)$$

Table 2.9 Process monitoring data for applying SRGM's

Project no	Contact review (X_1)		Development planning review (X_2)		Design completion review (X_3)		Test planning review (X_4)		Test completion review (X_5)	
	Number of detected faults	Total days for desolving faults	Number of detected faults	Total days for desolving faults	Number of detected faults	Total days for desolving faults	Number of detected faults	Total days for desolving faults	Number of detected faults	Total days for desolving faults
1	6	184	12	223	3	109	4	49	4	132
2	3	75	6	97	0	0	1	7	1	5
3	4	26	2	14	0	0	2	47	0	0
4	2	51	2	14	0	0	1	8	0	0
5	2	41	6	158	4	39	1	6	1	5
6	5	36	4	122	0	0	1	27	1	5
7	1	7	0	0	0	0	0	0	0	0
8	3	12	9	188	0	0	3	20	0	0
9	3	42	7	161	1	21	4	43	2	25
10	4	4	3	15	1	24	3	3	4	4
11	2	15	3	15	1	20	4	8	1	18
12	5	27	5	40	1	20	6	30	1	18
13	6	32	5	51	1	20	6	33	1	18
14	3	15	4	25	1	20	4	22	1	18
15	2	13	2	20	1	18	3	12	0	0
16	6	107	4	104	1	19	2	39	0	0
17	3	12	5	100	1	20	2	22	1	6
18	2	30	3	42	1	22	0	0	1	6
19	1	56	1	2	1	18	0	0	0	0
20	3	54	3	6	3	20	0	0	0	0
21	1	1	4	8	0	0	0	0	0	0

where E[A] represents the expected value for random variable A. And an instantaneous MTBF (mean time between software faults) is formulated as

$$MTBF_I(t) = \frac{1}{dH(t)/dt},$$ (2.52)

which is one of the substitute measures of the MTBF for the NHPP model.

Further, a software reliability function represents the probability that a software
failure does not occur in the time-interval $(t, t + x]$ $(t \geq 0, x \geq 0)$ given that the
testing or the user operation has been going up to time t. Then, if the counting process
$\{N(t), t \geq 0\}$ follows the NHPP with mean value function $H(t)$, the software
reliability function is derived as

$$R(x \mid t) = \exp[-\{H(t + x) - H(t)\}].$$ (2.53)

We have found that the logarithmic Poisson execution time model shows the
best goodness-of-fit in Projects 11–14 in which the test completion review's missing
values are complemented by collaborative filtering. We have also found that the
delayed S-shaped SRGM shows suitability in all projects. Therefore, if we select the
process monitoring progress ratio as the unit of testing-time for SRGM's based on an
NHPP, then the delayed S-shaped SRGM becomes very useful one for quantitative
software reliability assessment based on the process data derived from software
process monitoring activities.

Further, we show numerical illustration of software reliability assessment by using
the delayed S-shaped SRGM for Project 1. Figure 2.17 shows the estimated mean
value function and its 95 % confidence limits, where the parameter estimates are

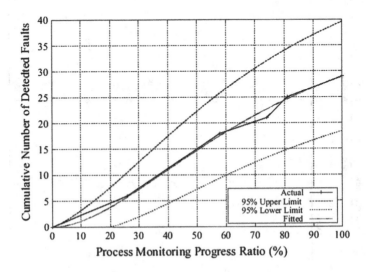

Fig. 2.17 The estimated mean value function for Project 1

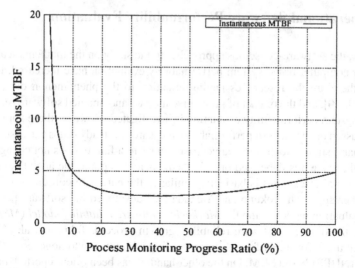

Fig. 2.18 The estimated instantaneous MTBF for Project 1

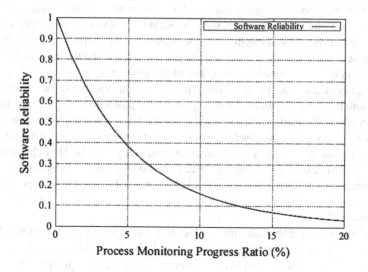

Fig. 2.19 The estimated software reliability function for Project 1

obtained as $\hat{a} = 39.67$ and $\hat{b} = 0.0259$. We can find that there are 10 remaining faults at the end of test completion review phase. Figure 2.18 shows the estimated instantaneous MTBF in (2.52). From Fig. 2.18, we can estimate the instantaneous MTBF at the finishing test completion review phase to be about 5 days. Figure 2.19 shows the estimated software reliability at process monitoring progress ratio $t = 100 (\%)$. From Fig. 2.19, if the process monitoring progress ratio is 120 %, we can find that a software failure will occur with high probability.

2.6 Operational Software Performability Evaluation

Recently, the software-conscious approaches to the study on the mathematical modeling for computer-based system performability evaluation have increased. For example, there are the researches paying attention to the phenomenon of software aging [39, 40], and the impact of preventive maintenance-named software rejuvenation on system performability evaluation is investigated. However these approaches are discussed on the basis of performability measures in steady states and assume that the probabilistic or stochastic characteristic in system failure does not change even though the system is debugged or refreshed. As to this point, the analytical framework in the above studies is basically similar to the hardware-conscious approach. As another approach, Tokuno and Yamada [41] developed the software performability evaluation model, using the *Markovian software reliability model (MSRM)* to describe the dynamic software reliability growth process. Tokuno et al. [42] also discussed the software performability modeling with the nonhomogeneous Poisson process (NHPP)-based SRM. On the other hand, it has been often reported that software reliability characteristic emerging in the field operation after release are quite different from the original predictions in the testing phase of the software development process, which are conducted with the SRMs [43, 44]. One of the main causes of this prediction gap is thought to be the underlying assumption for constructing SRMs that the software failure-occurrence phenomenon in the testing phase is similar to that in the user operation phase. Therefore, there exist some negative opinions about the above assumption, and several studies on field-oriented software reliability assessment have been conducted. Okamura et al. [45], Morita et al. [46], and Tokuno and Yamada [47] discussed the operation-oriented software reliability assessment models by introducing the concept of the accelerated life testing model which is often applied to the reliability assessment for hardware products. They characterized the difference between the testing and field-operation environments by assuming that the software failure time intervals are different generally from the viewpoint of the severities of both usage conditions. They have called the proportional constant the *environmental factor*.

However, the practical and reasonable estimation of the environmental factor remains the outstanding problem even in [45–47]. Originally it is impossible to apply the usual procedure for estimating the environmental factor since the software failures data in the operation phase observed from the software system in question are never obtained in advance. Accordingly, in the present circumstance, we have to decide the value of the environmental factor empirically and subjectively based on the similar software systems developed before. On the other hand, Pham [48, 49] presented the new mathematical reliability function, called *systemability* . The systemability function is defined as the probability that the system will perform its intended function for a specified mission time subject to the uncertainty of the operating environment. References [48, 49] assume that the hazard rate function in the field operation is proportional to one in the controlled in-house testing phase, and then considers the variation of the field environment by regarding the environmental factor as a random variable. As to the case of the software system, the operating environment in the testing phase is well-controlled one with less variation compared with the field operation environment [50]. In other words, the field operation environment includes the more uncertain factors than the testing environment in terms of the pattern of the execution load, the compatibility between the software system and the hardware platform, the operational profile, and so on [51]. It is almost impossible to prepare the test cases assuming all possible external disturbances. Therefore, the consideration of the randomness of the environmental factor can not only describe the actual operating environment more faithfully, but also reflect the subjective value of the environmental factor to the field operational software performability evaluation with a certain level of rationality.

In this section, we expand the meaning of systemability from the definition of Pham [49] into the system reliability characteristic considering the uncertainty and the variability of the field operating environment. Then we discuss the stochastic modeling for operational software performability measurement with systemability. We use the Markovian imperfect debugging model [52, 53] to describe the software reliability growth phenomena in the testing and the operation phases as the based model. We assume that the software system can process multiple tasks simultaneously and that the task arrival process follows an NHPP [41]. The stochastic behavior of the number of tasks whose processes can be completed within the processing time limit is modeled with the infinite-server queueing model [54]. Since performability is one of the operational-oriented characteristics, the consideration of systemability is meaningful in the system performability evaluation.

2.6.1 Markovian Software Reliability Model

2.6.1.1 Model Description

The following is the assumption for the generalized MSRM:

(A1) The debugging activity is performed as soon as the software failure occurs. The debugging activity for the fault having caused the corresponding software failure succeeds with the perfect debugging probability a ($0 \leq a \leq 1$), and fails with probability $b(= 1 - a)$. One perfect debugging activity corrects and removes one fault from the system and improves software reliability.

(A2) When n faults have been corrected, the next software-failure time-interval, U_n, follows the exponential distribution with the hazard rate λ_n which is denoted as $F_{U_n}(t) \equiv \Pr\{U_n \leq t\} = 1 - e^{-\lambda_n t}$. λ_n is a non-increasing function of n.

(A3) The debugging time of a fault is not considered.

Let $\{W(t), t \geq 0\}$ be a counting process representing the cumulative number of faults corrected up to the time t. Then $W(t)$ forms the Markov process whose state transition probability is given by perfect debugging probability a. From assumption A1, when i faults have already been corrected, after the next software failure occurs,

Fig. 2.20 State transition
diagram of $W(t)$

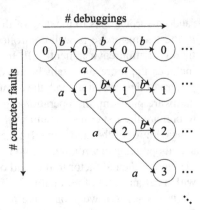

$$W(t) = \begin{cases} i+1 & \text{(in case of perfect debugging with probability } a) \\ i & \text{(in case of imperfect debugging with probability } b) \end{cases} . \quad (2.54)$$

Let $Q_{i,j}(t)$ $(i, j = 0, 1, 2, \ldots)$ denote the one-step transition probability which represents the probability that after making a transition into state i, the process $\{W(t), t \geq 0\}$ next makes a transition into state j (possibility $i = j$) in an amount of time less than or equal to t (see Osaki [55] for the more strict definition of $Q_{i,j}(t)$). The expressions for $Q_{i,j}(t)$'s are given by

$$Q_{i,i+1}(t) \equiv \Pr\{U_i \leq t, W(U_i) = i+1 | W(0) = i\}$$
$$= a(1 - e^{-\lambda_i t}), \quad (2.55)$$
$$Q_{i,i}(t) \equiv \Pr\{U_i \leq t, W(U_i) = i | W(0) = i\}$$
$$= b(1 - e^{-\lambda_i t}), \quad (2.56)$$

respectively. Figure 2.20 illustrates the state transition diagram of $W(t)$, where the integer in the circle denotes the number of corrected faults (i.e. the state of $W(t)$).

2.6.1.2 Distribution of Transition Time of $W(t)$

Let $S_{i,n}$ $(i \leq n)$ be the random variable representing the first passage time of $W(t)$ from state i to state n, and $G_{i,n}(t) \equiv \Pr\{S_{i,n} \leq t\}$ be the distribution function of $S_{i,n}$. First we consider the distribution function $G_{i,i+1}(t)$ and have the following renewal equation of $G_{i,i+1}(t)$:

$$G_{i,i+1}(t) = Q_{i,i+1}(t) + Q_{i,i} * G_{i,i+1}(t) \quad (i = 0, 1, 2, \ldots), \quad (2.57)$$

where $*$ denotes the Stieltjes convolution and $G_{i,i}(t) \equiv 1(t)$ (the unit function, $i = 0, 1, 2, \ldots$). From (2.55) and (2.56), the Laplace-Stieltjes (L-S) transform [55] of $G_{i,i+1}(t)$ is obtained as

$$\widetilde{G}_{i,i+1}(s) = \frac{a\lambda_i}{s + a\lambda_i} \quad (i = 0, 1, 2, \ldots), \tag{2.58}$$

where the L-S transform of $G_{i,i+1}(t)$ is defined as

$$\widetilde{G}_{i,i+1}(s) \equiv \int_0^\infty e^{-st} dG_{i,i+1}(t). \tag{2.59}$$

On the other hand, the following relationship between $S_{i,i+1}$ and $S_{i,n}$ holds:

$$S_{i,n} = \sum_{j=i}^{n-1} S_{j,j+1}. \tag{2.60}$$

Since $S_{i,i+1}$'s are mutually independent, $G_{i,n}(t)$ can be expressed as

$$G_{i,n}(t) = G_{i,i+1} * G_{i+1,i+2} * \cdots * G_{n-1,n}(t). \tag{2.61}$$

Therefore, the L-S transform of $G_{i,n}(t)$ is obtained as

$$\widetilde{G}_{i,n}(s) = \prod_{j=i}^{n-1} \frac{a\lambda_j}{s + a\lambda_j}. \tag{2.62}$$

Inverting (2.62), we obtain the solution of $G_{i,n}(t)$ as

$$\left. \begin{array}{l} G_{i,n}(t) \equiv \Pr\{S_{i,n} \leq t\} \\ = \displaystyle\sum_{m=i}^{n-1} A_m^{i,n}(1 - e^{-a\lambda_m t}) \quad (t \geq 0; n = 0, 1, 2, \ldots; i \leq n) \\ \left(\begin{array}{ll} A_{n-1}^{n-1,n} \equiv 1 & (i = n-1) \\ A_m^{i,n} = \displaystyle\prod_{\substack{j=i \\ j \neq m}}^{n-1} \frac{1}{1 - \lambda_m/\lambda_j} & (i < n-1; m = i, i+1, \ldots, n-1) \end{array} \right) \end{array} \right\}, \tag{2.63}$$

(see Tokuno et al. [53] for the detail of the above analysis).

2.6.1.3 State Occupancy Probability

Since $\{W(t), t \geq 0\}$ is the counting process, the following equivalence relation holds:

$$\{S_{i,n} \le t\} \Longleftrightarrow \{W(t) \ge n | W(0) = i\} \quad (i, n = 0, 1, 2, \ldots; \ i \le n). \tag{2.64}$$

Therefore, the following equation is obtained:

$$\Pr\{S_{i,n} \le t\} = \Pr\{W(t) \ge n | W(0) = i\}. \tag{2.65}$$

The state occupancy probability that the system is in state n at the time point t on the condition that the system was in state i at time point $t = 0$, $P_{i,n}(t)$, is given by

$$\begin{aligned}
P_{i,n}(t) &\equiv \Pr\{W(t) = n | W(0) = i\} \\
&= \Pr\{W(t) \ge n | W(0) = i\} - \Pr\{W(t) \ge n + 1 | W(0) = i\} \\
&= \Pr\{S_{i,n} \le t\} - \Pr\{S_{i,n+1} \le t\} \\
&= G_{i,n}(t) - G_{i,n+1}(t). \tag{2.66}
\end{aligned}$$

2.6.2 Consideration of Systemability

Hereafter, we discriminate the time domains between the testing and the operation phase. Let the superscript ^{O} and no superscript refer to the operation phase and the testing phase, respectively. Then, we assume the following relationship between U_i and U_i^O:

$$U_i = \alpha U_i^O \quad (\alpha > 0), \tag{2.67}$$

where α is called the *environmental factor*. From the viewpoint of the software reliability assessment, $0 < \alpha < 1$ ($\alpha > 1$) means that the testing phase is severer (milder) in the usage condition than the operation phase, and $\alpha = 1$ means that the testing environment is equivalent to the operational one. Then the distribution of U_i^O in the case where the environmental factor is treated as a constant is given by

$$\begin{aligned}
F_{U_i^O}(t|\alpha) &\equiv \Pr\{U_i^O \le t\} = \Pr\{U_i \le \alpha t\} \\
&= 1 - e^{-\alpha \lambda_i t}. \tag{2.68}
\end{aligned}$$

Based on (2.68), we obtain the distribution function of $S_{i,n}^O$ as

$$\begin{aligned}
G_{i,n}^O(t|\alpha) &\equiv \Pr\{S_{i,n}^O \le t\} \\
&= \sum_{m=i}^{n-1} A_m^{i,n} (1 - e^{-\alpha a \lambda_m t}), \tag{2.69}
\end{aligned}$$

where we should note that the coefficient $A_m^{i,n}$ in (2.69) is equivalent to $A_m^{i,n}$ in (2.63) (i.e. identical to the testing phase), and has no bearing on the environmental factor.

In general, the actual operating environment in the field is quite different from the controlled testing environment, and it is natural to consider that the external factors affecting software reliability characteristics fluctuate. Therefore, it is not appropriate that the environmental factor, α, introduced to bridge the gap between the software failure characteristics in the testing and the operation phases, is constant. Hereafter, we treat α as a random variable [48–50]. In this section, we consider the following two cases: the first is the model whose α follows the gamma distribution (G-model) and the second is the beta distribution (B-model).

2.6.2.1 G-Model

The G-model is assumed that the environmental factor α follows the gamma distribution whose density function is denoted as

$$f_\alpha(x) \equiv f_\alpha^G(x) = \frac{\theta^\eta \cdot x^{\eta-1} \cdot e^{-\theta x}}{\Gamma(\eta)}$$
$$(x \geq 0; \theta > 0, \eta \geq 1), \tag{2.70}$$

where $\Gamma(\eta) \equiv \int_0^\infty x^{\eta-1} e^{-x} dx$ is the gamma function, and θ and η are called the scale and the shape parameters, respectively. The G-model can be used to evaluate and predict the software reliability characteristic in the operation phase where the usage condition is severer than $(\alpha > 1)$, equivalent to $(\alpha = 1)$, or milder than $(0 < \alpha < 1)$ the testing environment. Then the mean and variance of α are given by

$$\mathrm{E}[\alpha] = \frac{\eta}{\theta}, \quad \mathrm{Var}[\alpha] = \frac{\eta}{\theta^2}, \tag{2.71}$$

respectively.

The posterior distribution of U_i^O is given by

$$F_{U_i^O}^G(t) = \int_0^\infty F_{U_i^O}(t|x) f_\alpha^G(x) dx = 1 - \frac{1}{(\lambda_i t/\theta + 1)^\eta}, \tag{2.72}$$

which is called the Pareto distribution. Assuming that the relationship between $S_{i,n}$ and $S_{i,n}^O$ can be described as $S_{i,n} = \alpha S_{i,n}^O$ approximately, we can obtain the posterior distribution of $S_{i,n}^O$ as

$$G_{i,n}^{OG}(t) = \int_0^\infty G_{i,n}^O(t|x) f_\alpha^G(x) dx$$
$$= \sum_{m=i}^{n-1} A_m^{i,n} \left[1 - \frac{1}{(a\lambda_m t/\theta + 1)^\eta} \right]. \tag{2.73}$$

2.6.2.2 B-Model

The B-model is assumed that the environmental factor follows the beta distribution whose density function is denoted as

$$f_\alpha(x) \equiv f_\alpha^B(x) = \frac{x^{\beta_1-1}(1-x)^{\beta_2-1}}{B(\beta_1, \beta_2)}$$
$$(0 < x < 1; \beta_1 > 0, \beta_2 > 0), \tag{2.74}$$

where β_1 and β_2 are the shape parameters, and $B(\beta_1, \beta_2) \equiv \int_0^1 t^{\beta_1-1}(1-t)^{\beta_2-1}dt = \Gamma(\beta_1)\Gamma(\beta_2)/\Gamma(\beta_1 + \beta_2)$ is the beta function. A beta-distribution random variable ranges between 0 and 1. Therefore, the B-model is appropriate to describe the operational software reliability characteristic only where the usage condition is estimated to be milder than the testing environment. Then the mean and the variance of α are given by

$$E[\alpha] = \frac{\beta_1}{\beta_1 + \beta_2}, \quad Var[\alpha] = \frac{\beta_1\beta_2}{(\beta_1 + \beta_2)^2(\beta_1 + \beta_2 + 1)}, \tag{2.75}$$

respectively.

The posterior distribution of U_i^O based on the B-model is given by

$$F_{U_i^O}^B(x) = \int_0^1 F_{U_i^O}(t|x) f_\alpha^B(x)dx$$
$$= 1 - e^{-\lambda_i t} \cdot M(\beta_2, \beta_1 + \beta_2; \lambda_i t), \tag{2.76}$$

where

$$M(c_1, c_2; x) \equiv \sum_{k=0}^{\infty} \frac{(c_1)_k}{(c_2)_k k!} x^k \quad ((c_1)_k \equiv \Gamma(c_1 + k)/\Gamma(k)), \tag{2.77}$$

is the Kummer function which is a kind of confluent hypergeometric functions [56]. Treating the relationship between $S_{i,n}$ and $S_{i,n}^O$ in the same way as the case of the G-model, we can obtain the posterior distribution of $S_{i,n}^O$ as

$$G_{i,n}^{OB}(t) = \int_0^1 G_{i,n}^O(t|x) f_\alpha^B(x)dx$$
$$= \sum_{m=i}^{n-1} A_m^{i,n}[1 - e^{-a\lambda_m t} \cdot M(\beta_2, \beta_1 + \beta_2; a\lambda_m t)]. \tag{2.78}$$

2.6.3 *Model Description and Analysis for Task Processing*

We make the following assumptions for system's task processing:

(B1) The number of tasks the system can process simultaneously is sufficiently large.

(B2) The process $\{N(t), \ t \geq 0\}$ representing the number of tasks arriving at the system up to the time t follows the NHPP with the arrival rate $\omega(t)$ and the mean value function $\Omega(t) \equiv \mathrm{E}[N(t)] = \int_0^t \omega(x)\mathrm{d}x$.

(B3) Each task has a processing time limit, T_r, which follows a general distribution whose distribution function is denoted as $F_{T_r}(t) \equiv \mathrm{Pr}\{T_r \leq t\}$.

(B4) The processing times of a task, Y is distributed generally; its distribution function is denoted as $F_Y(t) \equiv \mathrm{Pr}\{Y \leq t\}$. Each of the processing times is independent.

(B5) When the system causes a software failure in task processing or the processing times of tasks exceed the processing time limit, the corresponding tasks are canceled.

Figure 2.21 illustrates the relationship among the task completion/cancellation and the random variables U_n^O, T_r, and Y. Hereafter, we set the time origin $t = 0$ at the time point when the debugging activity is completed and i faults are corrected.

Let $\{X_i^1(t), t \geq 0\}$ be the stochastic process representing the cumulative number of tasks whose processes can be completed within the processing time limit out of the

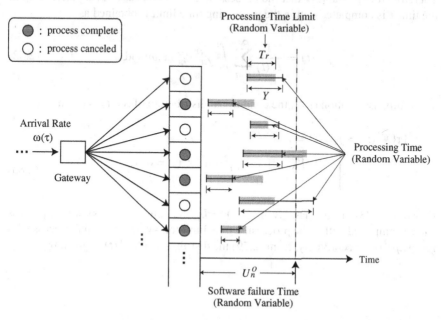

Fig. 2.21 Relationship between task completion/cancellation and U_n^O, T_r, Y

tasks arriving up to the time t. By conditioning with $\{N(t) = k\}$ $(k = 0, 1, 2, \ldots)$, we obtain the probability mass function of $X_i^1(t)$ as

$$\Pr\{X_i^1(t) = j\} = \sum_{k=0}^{\infty} \Pr\{X_i^1(t) = j | N(t) = k\} \times e^{-\Omega(t)} \frac{[\Omega(t)]^k}{k!}$$

$$(j = 0, 1, 2, \ldots). \quad (2.79)$$

From Fig. 2.21, the probability that the process of an arbitrary task is completed, given $\{W(t) = n\}$, is given by

$$\gamma_n^O \equiv \Pr\{Y < U_n^O, Y < T_r | W(t) = n\}$$

$$= \int_0^{\infty} \overline{F_{U_n^O}}(y) \overline{F_{T_r}}(y) \mathrm{d} F_Y(y), \quad (2.80)$$

where we denote $\overline{F}(\cdot) \equiv 1 - F(\cdot)$. Furthermore, from the property of the NHPP [54], given $\{N(t) = k\}$, the arrival time of an arbitrary task out of ones arriving up to the time t is the random variable having the following probability density function:

$$f(x) = \begin{cases} \dfrac{\omega(x)}{\Omega(t)} & (0 \le x \le t), \\ 0 & (x > t). \end{cases} \quad (2.81)$$

Therefore, the probability that the process of an arbitrary task having arrived up to the time t is completed within the processing time limit is obtained as

$$p_i^O(t) = \frac{1}{\Omega(t)} \sum_{n=i}^{\infty} \int_0^t \gamma_n^O P_{i,n}^O(x) \omega(x) \mathrm{d}x. \quad (2.82)$$

Then from assumption (B4), the conditional distribution of $X_i^1(t)$ is given by

$$\Pr\{X_i^1(t) = j | N(t) = k\}$$

$$= \begin{cases} \binom{k}{j} [p_i^O(t)]^j [1 - p_i^O(t)]^{k-j} & (j = 0, 1, 2, \ldots, k), \\ 0 & (j > k). \end{cases} \quad (2.83)$$

Equation (2.83) means that, given $\{N(t) = k\}$, the number of tasks whose process can be completed within the processing time limit follows the binomial process with mean $k p_i^O(t)$. Accordingly, from (2.79) the distribution of $X_i^1(t)$ is given by

$$\Pr\{X_i^1(t) = j\} = \sum_{k=j}^{\infty} \binom{k}{j} [p_i^O(t)]^j [1 - p_i^O(t)]^{k-j} e^{-\Omega(t)} \frac{[\Omega(t)]^k}{k!}$$

$$= e^{-\Omega(t)p_i^O(t)} \frac{[\Omega(t)p_i^O(t)]^j}{j!} \quad (j = 0, 1, 2, \ldots). \quad (2.84)$$

Equation (2.84) means that $\{X_i^1(t), t \geq 0\}$ follows the NHPP with the mean value function $\Omega(t)p_i^O(t)$.

Let $\{X_i^2(t), t \geq 0\}$ be the stochastic process representing the cumulative number of canceled tasks. By applying a similar discussion on $X_i^1(t)$, we have the distribution of $X_i^2(t)$ as

$$\left.\begin{aligned} \Pr\{X_i^2(t) = j\} &= e^{-\Omega(t)q_i^O(t)} \frac{[\Omega(t)q_i^O(t)]^j}{j!} \\ q_i^O(t) &= 1 - p_i^O(t) \end{aligned}\right\}. \quad (2.85)$$

Equation (2.85) means that $\{X_i^2(t), t \geq 0\}$ follows the NHPP with the mean value function $\Omega(t)q_i^O(t)$.

2.6.4 Derivation of Software Performability Measures

The expected number of tasks completable out of the tasks arriving up to the time t is given by

$$\Lambda_i^O(t) \equiv E[X_i^1(t)] = \sum_{n=i}^{\infty} \int_0^t \gamma_n^O P_{i,n}^O(x)\omega(x)dx. \quad (2.86)$$

Furthermore, the instantaneous task completion ratio is obtained as

$$\mu_i^O(t) \equiv \frac{d\Lambda_i^O(t)}{dt} \bigg/ \omega(t) = \sum_{n=i}^{\infty} \gamma_n^O P_{i,n}^O(t), \quad (2.87)$$

which represents the ratio of the number of tasks completed within the processing time limit to one arriving at the system per unit time at the time point t. We should note that (2.87) has no bearing on $\Omega(t)$, i.e. the task arrival process. As to $p_i^O(t)$ in (2.82), we can give the following interpretations:

$$p_i^O(t) = \frac{E[X_i^1(t)]}{E[N(t)]}. \quad (2.88)$$

That is, $p_i^O(t)$ is the cumulative task completion ratio up to the time t which represents the expected proportion of the cumulative number of tasks completed to one arriving at the system in the time interval $(0, t]$.

We should note that it is too difficult to use (2.86)–(2.88) practically since this model assumes the imperfect debugging environment and the initial condition i appearing in the above equations, which represents the cumulative number of faults corrected at time point $t = 0$, cannot be observed immediately. However, the numbers of software failures or debugging activities can be easily observed. Furthermore, the cumulative number of faults corrected when l debugging activities are performed, C_l, follows the binomial distribution whose probability mass function is given by

$$\Pr\{C_l = i\} = \binom{l}{i} a^i b^{l-i} \quad (i = 0, 1, 2, \ldots, l). \tag{2.89}$$

Accordingly, we can convert (2.86)–(2.88) into the functions of the number of debugging, l, i.e. we obtain

$$\Lambda^O(t, l) = \sum_{i=0}^{l} \binom{l}{i} a^i b^{l-i} \Lambda_i^O(t), \tag{2.90}$$

$$\mu^O(t, l) = \sum_{i=0}^{l} \binom{l}{i} a^i b^{l-i} \mu_i^O(t), \tag{2.91}$$

$$p^O(t, l) = \sum_{i=0}^{l} \binom{l}{i} a^i b^{l-i} p_i^O(t), \tag{2.92}$$

respectively. Equations (2.90)–(2.92) represent the expected cumulative number of tasks completable, the instantaneous and the cumulative task completion ratios at the time point t, given that the lth debugging was completed at time point $t = 0$, respectively.

2.6.4.1 In Case of G-Model

By substituting $G_{i,n}^{OG}(t)$ in (2.73) into $G_{i,n}^O(t)$ in (2.90)–(2.92), we can obtain the cumulative number of tasks completable, the instantaneous and the cumulative task completion ratios at the time point t as

$$\Lambda^{OG}(t, l) = \sum_{i=0}^{l} \binom{l}{i} a^i b^{l-i} \sum_{n=i}^{\infty} \int_0^t \gamma_n^{OG} P_{i,n}^{OG}(x) \omega(x) dx, \tag{2.93}$$

$$\mu^{OG}(t, l) = \sum_{i=0}^{l} \binom{l}{i} a^i b^{l-i} \sum_{n=i}^{\infty} \gamma_n^{OG} P_{i,n}^{OG}(t), \tag{2.94}$$

$$p^{OG}(t,l) = \frac{1}{\Omega(t)} \sum_{i=0}^{l} \binom{l}{i} a^i b^{l-i} \sum_{n=i}^{\infty} \int_0^t \gamma_n^{OG} P_{i,n}^{OG}(x)\omega(x)\mathrm{d}x, \qquad (2.95)$$

respectively, where

$$P_{i,n}^{OG}(t) = G_{i,n}^{OG}(t) - G_{i,n+1}^{OG}(t), \qquad (2.96)$$

$$\gamma_n^{OG} = \int_0^{\infty} \frac{1}{(\lambda_n y/\theta + 1)^n} \overline{F_{T_r}}(y)\mathrm{d}F_Y(y). \qquad (2.97)$$

2.6.4.2 In Case of B-Model

By substituting $G_{i,n}^{OB}(t)$ in (2.78) into $G_{i,n}^{O}(t)$ in (2.90)–(2.92), we can obtain the cumulative number of tasks completable, the instantaneous and the cumulative task completion ratios at the time point t as

$$\Lambda^{OB}(t,l) = \sum_{i=0}^{l} \binom{l}{i} a^i b^{l-i} \sum_{n=i}^{\infty} \int_0^t \gamma_n^{OB} P_{i,n}^{OB}(x)\omega(x)\mathrm{d}x, \qquad (2.98)$$

$$\mu^{OB}(t,l) = \sum_{i=0}^{l} \binom{l}{i} a^i b^{l-i} \sum_{n=i}^{\infty} \gamma_n^{OB} P_{i,n}^{OB}(t), \qquad (2.99)$$

$$p^{OB}(t,l) = \frac{1}{\Omega(t)} \sum_{i=0}^{l} \binom{l}{i} a^i b^{l-i} \sum_{n=i}^{\infty} \int_0^t \gamma_n^{OB} P_{i,n}^{OB}(x)\omega(x)\mathrm{d}x, \qquad (2.100)$$

respectively, where

$$P_{i,n}^{OB}(t) = G_{i,n}^{OB}(t) - G_{i,n+1}^{OB}(t), \qquad (2.101)$$

$$\gamma_n^{OB} = \int_0^{\infty} e^{-\lambda_n y} \cdot \mathrm{M}(\beta_2, \beta_1 + \beta_2; \lambda_n y)\overline{F_{T_r}}(y)\mathrm{d}F_Y(y). \qquad (2.102)$$

2.6.5 Numerical Examples

We present several numerical examples on system performability analysis [57] based on the above measures. Here we apply $\lambda_n \equiv Dc^n$ ($D > 0$, $0 < c < 1$) to the hazard rate [58]. For the distributions of the processing times, $F_Y(t)$, and the processing time limit, $F_{T_r}(t)$, we apply the gamma distribution with the shape parameter of two denoted by

$$F_I(t) \equiv H(t|\sigma_I) = 1 - (1 + \sigma_I t)e^{-\sigma_I t}$$
$$(t \geq 0; \ \sigma_I > 0; \ I \in \{Y, T_r\}). \qquad (2.103)$$

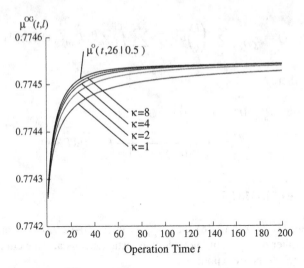

Fig. 2.22 Dependence of $\mu^{OG}(t, l)$ in G-model on κ in case of $E[\alpha] = 0.5$ ($a = 0.9, l = 26$; $D = 0.202, c = 0.950, \sigma_Y = 900, \sigma_{T_r} = 400$)

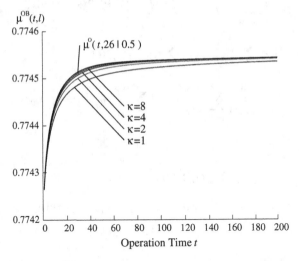

Fig. 2.23 Dependence of $\mu^{OB}(t, l)$ in B-model on κ in case of $E[\alpha] = 0.5$ ($a = 0.9, l = 26$; $D = 0.202, c = 0.950, \sigma_Y = 900, \sigma_{T_r} = 400$)

Figure 2.22 shows the dependence of the instantaneous task completion ratio, $\mu^{OG}(t, l)$ in G-model, in (2.94) on the value of κ, along with $\mu^{O}(t, l|0.5)$, which designates the instantaneous task completion ratio in the case where α is treated as a constant and $\alpha = 0.5$, where we set $\theta = \kappa\theta_0$, $\eta = \kappa\eta_0$; and $\theta_0 = 2.0$, $\eta_0 = 1.0$. Similarly, Fig. 2.23 shows the dependence of $\mu^{OB}(t, l)$ in B-model, in (2.99) on κ, where we set $\beta_1 = \kappa\beta_{10}$, $\beta_2 = \kappa\beta_{20}$; and $\beta_{10} = 1.0$, $\beta_{20} = 1.0$. In any value of

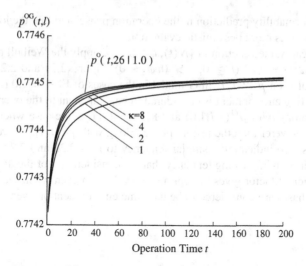

Fig. 2.24 Dependence of $p^{OG}(t, l)$ in G-model on κ in case of $E[\alpha] = 1.0$ ($a = 0.9, l = 26$; $D = 0.202, c = 0.950, \xi = 1.0, \phi = 2.0, \sigma_Y = 900, \sigma_{T_r} = 400, \theta_0 = 1.0, \eta_0 = 1.0$)

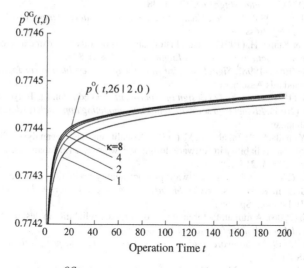

Fig. 2.25 Dependence of $p^{OG}(t, l)$ in G-model on κ in case of $E[\alpha] = 2.0$ ($a = 0.9, l = 26$; $D = 0.202, c = 0.950, \xi = 1.0, \phi = 2.0, \sigma_Y = 900, \sigma_{T_r} = 400, \theta_0 = 1.0, \eta_0 = 2.0$)

κ, $E[\alpha] = \eta_0/\theta_0 = \beta_{10}/(\beta_{10} + \beta_{20}) = 0.5$ is constant in both models, whereas $Var[\alpha] = \eta_0/(\kappa\theta_0^2)$ in G-model and $Var[\alpha] = \beta_{10}\beta_{20}/[(\beta_{10}+\beta_{20})^2(\kappa\beta_{10}+\kappa\beta_{20}+ 1)]$ in B-model, i.e. $Var[\alpha]$ is set as a decreasing function of κ in both models. From these setting, the larger value of κ means that the degree of conviction in terms of information on the environment factor is higher. These figures tell us that the higher degree of conviction of prior information on the environmental factor brings in more accurate performability prediction in the operation phase, and that the lower degree of conviction gives more pessimistic evaluation.

For the mean value function of $\{N(t), t \geq 0\}$, we apply the Weibull process, i.e. $E[N(t)] \equiv \Omega(t) = \xi t^{\phi}$ ($t \geq 0; \xi > 0, \phi > 0$). Figures 2.24 and 2.25 show the dependence of κ on $p^{OG}(t, l)$ in G-model in the cases of $E[\alpha] = 1.0$ (i.e. the case where the testing environment is considered to be equivalent to the operational one on average) along with $p^{O}(t, l | 1.0)$ and $E[\alpha] = 2.0$ (i.e. the case where the operation phase is severer than the testing phase) along with $p^{O}(t, l | 2.0)$, respectively. These figures also indicate the similar tendency to Figs. 2.22 and 2.23. Especially in Fig. 2.24, it is an interesting tendency that the consideration of the uncertainty of the environmental factor gives the optimistic evaluation even when the testing and pessimistic phases are considered to be the same environment on average.

References

1. Basili, V. R., & Reiter, R. W, Jr. (1979). An investigation of human factors in software development. *IEEE Computer Magazine, 12*, 21–38.
2. Curtis, B. (Ed.). (1985). *Tutorial: Human factors in software development.* Los Alamitos: IEEE Computer Society Press.
3. Nakajo, T., & Kume, H. (1991). A case history analysis of software error cause-effect relationships. *IEEE Transactions on Software Engineering, 17*, 830–838.
4. Taguchi, G. (Ed.). (1998). *Signal-to-noise raito for quality evaluation (in Japanese).* Tokyo: Japanese Standards Association.
5. Taguchi, G. (1976). *A method of design of experiment* (2nd ed., Vol. 1). Tokyo: Maruzen.
6. Yamada, S. (2011). *Elements of software reliability -modeling approach (in Japanese).* Tokyo: Kyoritsu-Shuppan.
7. Esaki, K., Yamada, S., & Takahashi, M. (2001). A quality engineering analysis of human factors affecting software reliability in software design review process (in Japanese). *Transactions of IEICE Japan, J84-A*, 218–228.
8. Yamada, S. (2008). Early-stage software product quality prediction based on process measurement data. In K. B. Misra (Ed.), *Springer handbook of performability engineering* (pp. 1227–1237). London: Springer.
9. Yamada, S. (2006). A human factor analysis for software reliability in design-review process. *International Journal of Performability Engineering, 2*, 223–232.
10. Miyamoto, I. (1982). *Software engineering -current status and perspectives (in Japanese).* Tokyo: TBS Publishing.
11. Esaki, K., & Takahashi, M. (1997). A software design review on the relationship between human factors and software errors classified by seriousness (in Japanese). *Journal of Quality Engineering Forum, 5*, 30–37.
12. E-Soft Inc., Internet Research Reports. Available: http://www.securityspace.com/s_survey/data/index.html
13. Yamada, S. (2002). Software reliability models. In S. Osaki (Ed.), *Stochastic models in reliability and maintenance* (pp. 253–280). Berlin: Springer.
14. MacCormack, A., Rusnak, J., & Baldwin, C. Y. (2006). Exploring the structure of complex software designs: An empirical study of open source and proprietary code. *Informs Journal of Management Science, 52*, 1015–1030.
15. Kuk, G. (2006). Strategic interaction and knowledge sharing in the KDE developer mailing list. *Informs Journal of Management Science, 52*, 1031–1042.

16. Zhou, Y., & Davis, J. (2005). Open source software reliability model: An empirical approach. In *Proceedings of the Fifth Workshop on Open Source Software Engineering (WOSSE)* (pp. 67–72).

17. Li, P., Shaw, M., Herbsleb, J., Ray, B., & Santhanam, P. (2004). Empirical evaluation of defect projection models for widely-deployed production software systems. *Proceedings of the 12th International Symposium on Foundations of, Software Engineering (FSE-12)* (pp. 263–272).

18. Arnold, L. (1974). *Stochastic differential equations-theory and applications.* New York: John Wiley & Sons.

19. Wong, E. (1971). *Stochastic Processes in Information and Systems.* New York: McGraw-Hill.

20. Yamada, S., Kimura, M., Tanaka, H., & Osaki, S. (1994). Software reliability measurement and assessment with stochastic differential equations. *IEICE Transactions on Fundamentals of Electronics, Communications, and Computer Sciences, E77-A,* 109–116.

21. The Apache HTTP Server Project, The Apache Software Foundation. Available: http://httpd. apache.org/

22. Apache Tomcat, The Apache Software Foundation. Available: http://tomcat.apache.org/

23. PostgreSQL, PostgreSQL Global Development Group. Available: http://www.postgresql.org/

24. Tamura, Y., & Yamada, S. (2007). Software reliability growth model based on stochastic differential equations for open source software. *Proceedings of the 4th IEEE International Conference on Mechatronics, CD-ROM (ThM1-C-1).*

25. Tamura, Y., & Yamada, S. (2006). A flexible stochastic differential equation model in distributed development environment. *European Journal of Operational Research, 168,* 143–152.

26. Tamura, Y., & Yamada, S. (2009). Optimisation analysis for reliability assessment based on stochastic differential equation modeling for open source software. *International Journal of Systems Science, 40,* 429–438.

27. Tamura, Y., & Yamada, S. (2013). Reliability assessment based on hazard rate model for an embedded OSS porting phase. *Software Testing, Verification and Reliability, 23,* 77–88.

28. Satoh, D. (2000) A discrete Gompertz equation and a software reliability growth model. *IEICE Transactions on Information and Systems, E83-D,* 1508–1513.

29. Satoh, D., & Yamada, S. (2001). Discrete equations and software reliability growth models. *Proceedings of the 12th International Symposium on Software Reliability Engineering (IS-SRE'01)* (pp. 176–184).

30. Inoue, S., & Yamada, S. (2007). Generalized discrete software reliability modeling with effect of program size. *IEEE Transactions on System, Man, and Cybernetics (Part A), 37,* 170–179.

31. Hirota, R. (1979). Nonlinear partial difference equations. V. Nonlinear equations reducible to linear equations. *Journal of Physical Society of Japan, 46,* 312–319.

32. Bass, F. M. (1969). A new product growth model for consumer durables. *Management Science, 15,* 215–227.

33. Satoh, D. (2001). A discrete Bass model and its parameter estimation. *Journal of Operations Research Society of Japan, 44,* 1–18.

34. Kasuga, K., Fukushima, T., & Yamada, S. (2006). A practical approach software process monitoring activities (in Japanese). *Proceedings of the 25th JUSE Software Quality Symposium* (pp. 319–326).

35. Yamada, S., & Fukushima, T. (2007). *Quality-oriented software management (in Japanese).* Tokyo: Morikita-Shuppan.

36. Yamada, S., & Takahashi, M. (1993). *Introduction to software management model (in Japanese).* Tokyo: Kyoritsu-Shuppan.

37. Yamada, S., & Kawahara, A. (2009). Statistical analysis of process monitoring data for software process improvement. *International Journal of Reliability, Quality and Safety Engineering, 16,* 435–451.

38. Yamada, S., Yamashita, T., & Fukuta, A. (2010). Product quality prediction based on software process data with development-period estimation. *International Journal of Systems Assurance Engineering and Management, 1,* 69–73.

39. Pfening, A., Garg, S., Puliafito, A., Telek, M., & Trivedi, K. S. (1996). Optimal software rejuvenation for tolerating soft failures. *Performance Evaluation, 27–28,* 491–506.

40. Garg, S., Puliafito, A., Telek, M., & Trivedi, K. S. (1998). Analysis of preventive maintenance in transactions based software systems. *IEEE Transactions on Computers, 47*, 96–107.
41. Tokuno, K., & Yamada, S. (2008). Dynamic performance analysis for software system considering real-time property in case of NHPP task arrival. *Proceedings of 2nd International Conference on Secure System Integration and Reliability Improvement (SSIRI 2008)* (pp. 73–80).
42. Nagata, T., Tokuno, K., & Yamada, S. (2011). Stochastic performability evaluation based on NHPP reliability growth model. *International Journal of Reliability, Quality, and Safety Engineering, 18*, 431–444.
43. Jeske, D. R., Zhang, X., & Pham, L. (2005). Adjusting software failure rates that are estimated from test data. *IEEE Transactions on Reliability, 54*, 107–114.
44. Pham, H. (2006). *System software reliability*. London: Springer.
45. Okamura, H., Dohi, T., & Osaki, S. (2001). A reliability assessment method for software products in operational phase: Proposal of an accelerated life testing model. *Electronics and Communications in Japan, 84*, 25–33.
46. Morita, H., Tokuno, K., & Yamada, S. (2005). Markovian operational software reliability measurement based on accelerated life testing model. *Proceedings of the 11th ISSAT International Conference on Reliability and Quality in Design* (pp. 204–208).
47. Tokuno, K., & Yamada, S. (2007). User-oriented and -perceived software availability measurement and assessment with environmental factors. *Journal of Operations Research Society of Japan, 50*, 444–462.
48. Pham, H. (2005). A new generalized systemability model. *International Journal of Performability Engineering, 1*, 145–155.
49. Pham, H. (2010). Mathematical systemability function approximations. *Proceedings of the 16th ISSAT International Conference on Reliability and Quality in Design* (pp. 6–10).
50. Teng, X., & Pham, H. (2006). A new methodology for predicting software reliability in the random field environments. *IEEE Transactions on Reliability, 55*, 458–468.
51. Lyu, M. R. (Ed.). (1996). *Handbook of software reliability engineering*. Los Alamitos: McGraw-Hill, IEEE Computer Society Press.
52. Tokuno, K., & Yamada, S. (2000). An imperfect debugging model with two types of hazard rates for software reliability measurement and assessment. *Mathematical and Computer Modeling, 31*, 343–352.
53. Tokuno, K., Kodera, T., & Yamada, S. (2009). Generalized markovian software reliability modeling and its alternative calculation. *International Journal of Reliability, Quality and Safety Engineering, 16*, 385–402.
54. Ross, S. M. (2007). *Introduction to probability models* (9th ed.). San Diego: Academic Press.
55. Osaki, S. (1992). *Applied stochastic system modeling*. Heidelberg: Springer.
56. Oldham, K. B., Myland, J. C., & Spanier, J. (2008). *An atlas of functions, with equator, the atlas function calculator* (2nd ed.). New York: Springer.
57. Tokuno, K., Fukuda, T., & Yamada, S. (2012). Operational software performability evaluation based on markovian reliability growth model with systemability. *International Journal of Reliability, Quality and Safety, Engineering, 19*, 1240001.
58. Moranda, P. B. (1979). Event-altered rate models for general reliability analysis. *IEEE Transactions on Reliability, R-28*, 376–381.

Index

S. Yamada, *Software Reliability Modeling*, SpringerBriefs in Statistics,
DOI: 10.1007/978-4-431-54565-1, © The Author(s) 2014